Monographs on soil survey

General Editors
P. H. T. BECKETT
V. C. ROBERTSON
R. WEBSTER

Aerial photography
and remote sensing
for soil survey

L. P. WHITE

CLARENDON PRESS · OXFORD
1977

Oxford University Press, Walton Street, Oxford OX2 6DP

OXFORD LONDON GLASGOW NEW YORK
TORONTO MELBOURNE WELLINGTON CAPE TOWN
IBADAN NAIROBI DAR ES SALAAM LUSAKA ADDIS ABABA
KUALA LUMPUR SINGAPORE JAKARTA HONG KONG TOKYO
DELHI BOMBAY CALCUTTA MADRAS KARACHI

© Oxford University Press 1977

British Library Cataloguing in Publication Data
White, Leslie Paul
    Aerial photography and remote sensing
    for soil survey. – (Monography on soil survey)
    Index.
    ISBN 0-19-854509-6
    1. Title   2. Series
    631.4'7   S591
    Aerial photography in soil surveys

Set by Hope Services, Wantage
and Printed in Great Britain
by Morrison & Gibb Ltd., Edinburgh

Editors' Foreword

Robin Clarke's classic *The Study of the soil in the field* has guided field pedologists and soil surveyors for forty years. It still has something to say. His discussion of the identification of soils on air photographs remains unsurpassed.

Even so, no science remains static. As our understanding of the soil in the field has increased, the aims of soil survey and land evaluation have broadened and new techniques have been developed to achieve them. The subject is now too large to be covered in a small volume. So this is the first of a series of new handbooks. Some, on field techniques, are intended for reference in the field: some, like this one, are intended to inform the field man of new possibilities.

Handbooks on field records and soil descriptions, soil classification, quantitative methods, and land evaluation are also in preparation. The editors will appreciate suggestions for further titles.

P.H.T.B.
V.C.R.
R.W.

Preface

Aerial photographs have been widely employed in soil survey for many years and for many types of work they are a standard requirement. Even so, the advantages and disadvantages of using different types of photography in different applications often hinge on the technical limitations of the techniques of photography, which are not always appreciated by the user. To complicate the matter, remote sensing techniques using non-photographic imaging devices from aircraft and satellites for mapping and monitoring the surface of the earth are attracting increasing interest. The value of these new methods for soil mapping work has yet to be fully determined and the potential user will also require technical knowledge to appreciate the scope of their potential use.

This monograph is intended to describe the use of aerial photography for soil mapping and to put it in the context of remote sensing in general. It describes the working principles of cameras and other devices that obtain images within and beyond the visible spectrum, the ways in which the images are produced and processed, and how they are used in soil survey.

I should like to thank my former colleagues in Hunting Technical Services Ltd. and my new colleagues in General Technology Systems Ltd. for their advice and encouragement. I am also grateful to Hunting Surveys Ltd. for providing the photographic illustrations.

Hounslow   L.P.W.
June 1976

Contents

List of plates

1. Aerial photography and other remote sensing methods

The idea of compiling boundaries between different sorts of soil on aerial photographs in the first instance, rather than on conventional maps, is straightforward enough. A conventional planimetric map carries only a limited amount of information on natural, as opposed to cultural, features, and although details such as spot-heights and contours may be valuable in interpreting the natural features of the landscape, they cannot in themselves indicate soil properties directly. On the other hand, a photograph is a permanent record of landscape which can simulate the visual experience of the scene — albeit from only one position and in an incomplete or distorted form. As such it reveals the subtle and multifarious features of the natural land surface and its vegetation cover, some of which are related to the properties of the soils and the boundaries between soils.

Aerial photography has been with us for a long time. Even during the American Civil War photographs were taken from balloons for military reconnaissance, and during the First World War cameras were extensively used from aircraft. The possibilities of using aerial photography for systematic topographic mapping and for mapping natural resources were explored between the First and Second World Wars and photogrammetric techniques are now fully established. More recently photographs have been taken from space satellites, and new methods of recording pictures have been developed. These new techniques and photography are known jointly as *remote sensing* — a term that was not current before 1960. (It originated at the Willow Run Laboratories of the University of Michigan, U.S.A., now the Environmental Research Institute of Michigan.)

Remote sensing, though not precisely defined, includes all methods of obtaining pictures or other forms of electromagnetic records of the Earth's surface from a distance, and the treatment and processing of the picture data. The term has even been extended to some forms of sea-bed survey and atmospheric monitoring. It covers the use of sensing instruments — cameras and others — and the handling and interpretation of the photographs or other images and non-picture data that these produce. In practice, photography is used more than any other remote sensing technique and photographic film is used much more than any other medium to record picture data from other sensor systems such as scanners and side-looking radar. Not surprisingly the term 'remote sensing' has met with some resistance from users of conventional aerial photography and practitioners of photogrammetric mapping, who resent their established disciplines being absorbed by others (if only linguistically) using new techniques which are often still largely unproved in practical application.

Remote sensing then, in the widest sense, is concerned with detecting and recording electromagnetic radiation from target areas in the field of view of the sensor instrument. This radiation may have originated directly from separate components of the target area; it may be solar energy reflected from them; or it may be reflections of energy transmitted to the target area from the sensor itself.

Electromagnetic radiation is characterized by its frequency or by its wavelength (which is inversely proportional to the frequency), and by its intensity. The radiation from any target can be expressed as a spectrum in which the most familiar parts, from the short to longer wavelengths, are the ultraviolet, visible, infrared, and radio sectors (Fig. 1.1.).

Human eyes register passively reflected radiation and are sensitive to a 'visible' band between 380 and 780 nm (nanometers: 0·38 to 0·78 μm) wavelength. Colour vision provides a multi-spectral capability. Normal colour vision divides the visible band into the primary colour segments of blue, green, and red as in the familiar rainbow spectrum. Since our eyes are paired, with overlapping fields of view, they also generate a stereoscopic or three-dimensional image in the brain from which we derive perspective or perception of depth. Eyes are excellent devices for registering images but unlike cameras they are not capable of making permanent records. It is likely that the human eye is specifically sensitive to the 'visual' spectrum because it is in this band, with a peak at about 0·5 μm, that solar radiation is at its most intense and terrestrial objects have a high reflectance. If the quality of sunlight were different, or if the atmosphere had different absorption properties, the spectral sensitivity of our eyes might have evolved differently. For instance, if our sun were a blue dwarf or red giant star our 'visual' range might have extended further into the ultraviolet or infrared parts of the spectrum respectively.

Our ears are also remote sensing organs, though they are not capable of forming images. They operate in sonar bands of very long wavelength and act as directional and ranging devices as well as forming part of our personal communication system. Because water is opaque to most of the electromagnetic radiation that passes freely through the atmosphere, sonar vibrations are used in underwater mapping systems such as side-scan sonar.

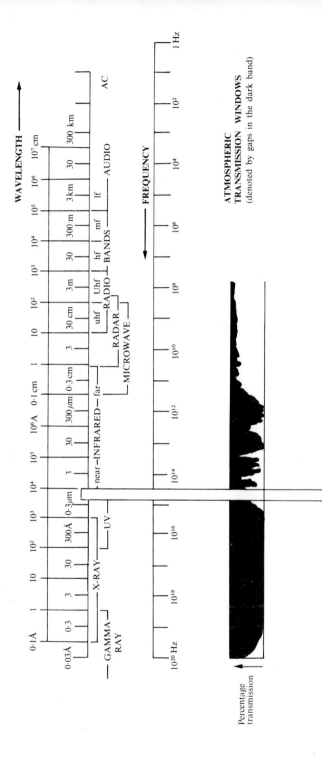

FIG. 1.1. The electromagnetic spectrum and atmospheric windows.

Whether received directly from the sun or from an alternative natural or artificial source, all radiation interacts with the components of the Earth's surface and is modified by them. It is reflected, transmitted, or absorbed in varying proportions, according to the nature of the surface and the wavelength and intensity of the radiation. The Earth's surface also emits its own radiation. According to Planck's law, any object at a temperature above absolute zero emits radiation and the peak of its emission spectrum moves progressively towards shorter wavelengths as the physical temperature of the object increases (Stefan's law). Thus a white-hot poker is hotter than a red-hot poker. For the general bulk of the materials making up the Earth's surface the peak of this self-emitted radiation is in the middle part of the infrared spectrum at about 10 μm.

By and large, remote sensing is concerned mainly with reflectances that originate from the immediate surface of the targets. It thus records only the superficial properties. Self-emitted radiation, on the other hand, may be directly influenced by conditions at depth in the target media.

The polarization of the radiation may also be affected by its interaction with the target. Polarization is 'the direction of the field vector', that is, the plane of propagation of the electric component of the electromagnetic radiation (the magnetic component propagates at right-angles to this). Radiation when received may be polarized in a different direction from that in which it was propagated. In some forms of radar, for instance, the signal is transmitted with a horizontal polarization, but reaction with the target induces a number of polarization directions in the backscatter (reflectance) and the receiver aerial is designed to receive only the vertically polarized component.

The atmosphere between an airborne or spaceborne sensor and its target is a selective filter for radiation and absorbs or scatters different fractions at different wavelengths. The use of sensors is thus confined to those parts of the electromagnetic spectrum that are least affected by the atmosphere. These are referred to as 'atmospheric windows' (Fig. 1.1).

The scattering of visible waveband radiation in a cloud-free atmosphere is due to the Rayleigh and Mei effects. The Rayleigh effect results from the interaction of light with the molecules of the atmospheric gases. It becomes progressively more marked towards the shorter wavelengths (the ultraviolet end of the spectrum), and the predominance of scattered blue light in the visible band gives the sky its characteristic blue colour. Mei scattering is due to larger, aerosol, particles in the atmosphere: dust, pollen, or water droplets. It causes most dispersion at the longer, red, end of the visible spectrum.

Each kind of sensor records radiation from a different sector of the electromagnetic spectrum. The spectral bands in which they operate are thus defined firstly by the sensitivity of the detecting elements or recording media of the sensor, and secondly by the atmospheric windows available to them. Sensor systems are also constrained by the amount of radiation available from the target. Although there is, for instance, a large amount of solar reflectance in the daytime, reflected visible and near-infrared radiation is drastically reduced at night and this limits the use of photography. Although all objects produce detectable levels of thermal infrared and microwave radiation, these may be too low for remote sensors to distinguish them satisfactorily from the general background radiation, commonly referred to as 'noise'. It is possible to solve this problem by illuminating the target in selected wavebands, as a flash-bulb is used with a camera. This is the principle of an 'active' as opposed to a 'passive' sensor system.

Passive sensor systems, such as cameras and most forms of scanner, merely record naturally reflected or self-emitted radiation. Active systems such as radars emit their own radiation signals and record their reflectance. Table 1.1. lists the principal kinds of sensor system and the spectral bands in which they operate. The most familiar of these, of course, are photographic cameras. We shall also be concerned with scanners, side-looking radar, and the television (vidicon) cameras used in space satellites.

Table 1.1.
Wavebands recorded by the principal remote sensing instruments

Waveband	Wavelength (μm)	Sensor System
Ultraviolet	0·3 — 0·38	Photographic cameras
Visible	0·38—00·78	Photographic cameras / Video cameras / Scanners
Near-infrared	0·78— 0·9	Photographic cameras / Video cameras / Scanners
	0·9 — 1·1	Video cameras / Scanners
Middle infrared	3·5 — 5·5	Scanners
Far infrared	8·0 —14·0	Scanners
Far infrared–microwave	14—1000	Scanners
Microwave–radio	1000 — 1 metre	Radar

As we have seen, photographic cameras are passive systems that record images in the visible part of the spectrum, and the immediately adjacent parts of the ultraviolet and infrared, on photosensitive film. They collect the image through a lens system and the amount of light received at each exposure is controlled by a shutter.

Television cameras are also passive instruments operating in much the same wavebands as photographic cameras. Their images, however, are formed as patterns of electrical charges on an image plate which is scanned by an electron beam and converted to electrical signals. These can be relayed over long distances by radio links or stored as videotape recordings for later replay and display.

Line-scanners are passive systems that employ a rotating optical system to scan successive strips of ground along the track of the aircraft or satellite. The radiation picked up along the lines making up a swath of coverage is focused on to detector elements. These elements, different types of which are sensitive to different parts of the spectrum, generate electronic signals which are converted to a pictorial record by electronic processing. According to the detectors used, scanners can record in a wide range of spectral wavebands from the ultraviolet to the microwave.

Both photographic and television (video) cameras record all of the picture or 'scene' focused on to the recording medium at the same time, in what is known as the 'image plane'. Scanner systems, on the other hand, build up the picture from spectral information as it is collected directly from the target area, that is, in the 'object plane'. Unlike cameras, the information is recorded continuously, without the use of shutters.

Like scanners, side-looking radar (SLR) systems also collect their information in the object plane rather than in the image plane. They are, however, active systems that generate an image from the returning signals (backscatter) from a pulsed radio beam. An image swath along the track of the aircraft is built up from the lines illuminated by successive pulses.

These main types of sensor system are described in Chapters 2, 5, 6, and 7.

Both colour and black and white (panchromatic) photography normally record the scene in the full visible waveband. The various components of the visible spectrum can, however, be recorded separately by using colour filters to divide the radiation to be recorded into its constituent components. Multi-spectral imaging techniques can in theory similarly be used with any type of sensor that has a wide-band capability. The main limitations are the sensitivity of the detecting elements or of the recording media and the efficiency of the filters. If the intensity of target radiation can be measured separately in a number of precisely defined spectral bands it becomes possible to describe the appearance of the target in mathematical terms or 'spectral signatures', which leads to the possibility of automatic target recognition and mapping. This is discussed in Chapter 8.

The radiation from a particular target may be modified by absorption or scattering by the atmosphere between it and the sensor, by the background radiation or 'noise', and inconsistencies in the sensor systems themselves. Imaging systems may therefore have to be supported by non-imaging instruments. Supplementary 'calibration' sensors can be used to measure accurately the radiation characteristics of samples of the target. The readings are then compared with the image records in the same wavebands in order to correct and quantify the image data.

In the visible and infrared bands calibrating instruments include single-band and multi-spectral radiometers and spectro-radiometers to record the total reflectance through a given spectrum from a single point. In the radio wavelengths the instruments are passive microwave recorders and active radar scatterometers. Calibrating instruments may be airborne with the imaging sensors or they may be used on the ground to provide the measurements known as 'ground truth' in support of aerial surveys.

With the exception of photographic cameras, all remote sensing imaging systems translate the electronic signals, which are transformations of the radiation received from their targets, into equivalent (analogous) light signals, which can be recorded on film to give photographic reproductions of the images. A term often used for these is 'hard copy'. Because film is used in this way to record transformed spectral images as well as for direct photographic recording, there is a tendency to call all remote sensing pictures photographs. Terms such as 'radar photograph' and 'infrared photograph' for what are microwave and thermal records are contradictions in terms. For this reason it is better to use the term 'imagery' for records that are not directly registered on to film by cameras, and to restrict 'photograph' to records in the visible and adjacent near-infrared bands made directly by cameras.

In any electronic, as opposed to a photographic, sensing system the initial form of image data is a continuously varying electrical signal or analog voltage. In a television system this can be directly and immediately displayed on a screen, giving what is known as a 'real-time' display. Alternatively the signal voltage can be recorded on magnetic tape, as for example videotape, and used later to generate a picture. As a variation of this, analogue information can be digitized at any stage to facilitate data-handling and analysis procedures.

To create 'hard copy' images or prints from analog or digital tapes, various film-writing systems are used. In these a light source is made to follow the same pattern across an unexposed film as that in which the original signals were recorded on the ground, while its intensity is modulated by variations in the recorded voltage. Various electron-beam, laser, and mechanical scanning systems that do this are discussed in later chapters.

Most electronic remote sensing systems thus have three basic modules:

(1) *Sensor system:* video camera; scanner; radar;
(2) *Tape recorder:* analog or digital recording;
(3) *Image processer:* film-writing system.

As the tape recorder is common to nearly all systems it follows that it should be possible to put data from any electronic sensor through a common image processer provided they are first converted to the appropriate format. Further, photographs obtained by conventional cameras may also be scanned to produce data in similar format, which can be selectively processed and enhanced. These techniques are described in Chapter 8.

Data collection by remote sensing systems requires aircraft or other airborne or space platforms on which to mount the equipment. The most familiar of these are the aerial survey aircraft operated by private or government survey organizations and the military. Balloons, kites, and rockets have also been used, though nearly always in experimental roles. Rockets have been used to carry cameras beyond the atmosphere, but the main vehicles for this region are space satellites, manned or unmanned.

In general, remote-sensing methods, other than conventional aerial photography, appear to offer only limited benefits to soil studies of a conventional kind. The sensors themselves have been developed for a variety of different purposes and their imagery may not be particularly suitable for soil survey. There have been a number of experimental soil and soil-related studies using advanced sensor systems and mapping techniques but few routine soil surveys are carried out using these in preference to conventional photographic interpretation.

Possible applications of new remote-sensing methods for soil surveys are (i) as an alternative to conventional aerial photography where this is not available and cannot be satisfactorily obtained; (ii) to obtain information in wavebands that are not recorded by photographic methods; (iii) to obtain more accurate spectral data than can be provided by photography and in forms that are directly compatible with automatic mapping techniques.

Much attention is being paid to the *Landsat* earth resources technology satellites which provide small-scale images with relatively good resolution. These are proving of real value to projects involving soil survey, particularly when they are used in combination with other survey techniques. Side-looking airborne radar coverage is also being used for practical projects. Generally, however, soil survey requires high-resolution stereoscopic imagery that is relatively correct planimetrically. This requirement is most readily and cheaply supplied by photographic survey cameras. Except perhaps for general reconnaissance, soil mapping also requires a high density of ground checking; so the need for alternative methods for collection and interpretation of complex spectral data for remote analysis is limited. Remote sensing by other than photographic methods has a great deal to offer in a number of other natural resources mappling applications, for instance in geology. For basic soil survey its use seems more limited, but the techniques can certainly be of value in such soil-related activities as monitoring moisture conditions and vegetation cover for irrigation and crop management.

2. Photography and photographic products

Photography is the most familiar form of remote sensing. It is also the most used, and despite some inherent disadvantages it is likely to remain so. It is easy to obtain and use; the processing necessary to produce a picture is relatively simple; and the film is an excellent medium on which to store data.

The principle of the camera is that light rays radiating from an object are refracted by a lens and made to converge to form an image in the focal plane of the lens. The focal length of a camera is the distance between the principal plane of the lens and its focal plane, that is from the optical centre of the lens to its principal focus on the image plane. The image is registered on photosensitive film in the focal plane (Fig. 2.1). Basically the amount of light, or luminous flux, that reaches the focal plane is controlled by the speed of opening and closing of a shutter behind the lens and by the size of the aperture in the shutter through which the light passes. On most cameras both shutter speed and aperture are adjustable to allow for different lighting conditions.

Photographic films consists of an emulsion of photosensitive silver halide grains in gelatine on a transparent plastic base. On exposure to light, some of the halide grains are partially reduced to silver, forming a 'latent image' of metallic nuclei. This latent image is not perceptible but it can be 'developed' by a reducing solution known as developer which acts on the previously reduced nuclei, enlarging them by converting more of the halide salts to metallic silver. When the developer has reduced enough silver to give the required image density, which is determined by both time and temperature, the image is 'fixed' by another solution which dissolves out all the unreduced halide. This leaves the developed image as a negative impression of the scene to which the film was exposed. The degree of darkness, or density, in any part of the negative corresponds to the relative brightness of that part of the real scene which it represents.

An image in positive form is obtained by exposing either another film or an emulsion-coated opaque paper to light which has passed through the negative. This 'printing' process reverses the tonal densities of the negative so that they correspond to those of the original scene. If the negative is printed by direct contact with the paper it gives a 'positive print' at the same scale. Alternatively it can be projected on to the copy material to provide varying degrees of enlargement.

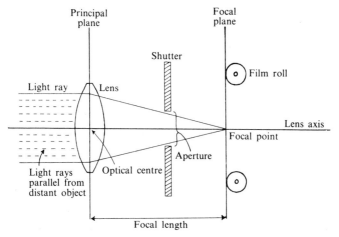

FIG. 2.1. The workings of a camera.

Halide grains are, in fact, sensitive only to ultraviolent and blue light; so to make them register the visible spectrum, the ultraviolet light must be filtered out and their sensitivity must be extended into the longer (red) end of the spectrum. The first is simple enough, for optical glass cuts out much of the ultraviolet radiation. It is, however, usually necessary to filter out more of the ultraviolet in order to compensate for differences between the sensitivity of film and that of the human eye in this range. The sensitivity of the film is extended into the red, (and infrared), parts of the spectrum by 'coupling' the halide grains with dyes that absorb the required wavelengths selectively and thus activate them; for instance, a green dye is used to increase the absorption of red light. A normal black and white panchromatic film may thus be sensitized from the violet (400 nm) to the red (650 nm), and beyond, into the near infrared (950 nm). Fig. 2.2 compares the sensitivities of several commonly used films.

As the amount of light falling on the film is increased the density of the resulting negative increases. Films differ in how much their densities increase with exposure, and a plot of negative density against the log of the exposure gives a curve that is unique for each kind of film (Fig. 2.3). This 'characteristic' or 'log e' curve is used to determine the latitude or range of exposure over which the density of the film increases in proportion to its exposure, and thereby achieves acceptable reproduction. The latitude is also related to one of the most important characteristics of a film, its *gamma* or contrast-rendering power, which is the slope of the log e curve. The 'speed' of a film is the reciprocal of the exposure necessary to bring the emulsion to a standard point on the characteristic

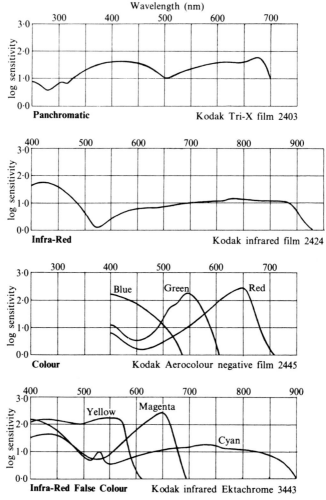

FIG. 2.2. Spectral sensitivity curves for aerial films. The curves represent a density of 1.0 above gross fog.

curve. Provided that the darkest tone on the image is not above this point, an acceptable photograph is produced.

The speed of a film depends on its grain size, which can vary from 0·1 to 5·0 μm. Large grains require a thick emulsion layer but coarse-grained films are more sensitive, or faster, than fine-grained films. They also have lower resolving power since the grain size imposes the ultimate limit on the minimum size of object that can be recorded on the negative. Films used for aerial survey must have high resolution, and typically they have grain sizes of about $1 \cdot 0 \, \mu m$.

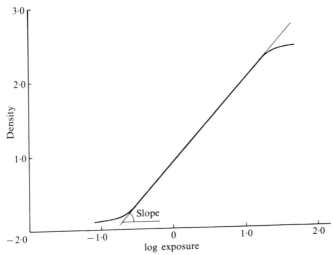

FIG. 2.3. Log e or characteristic curve of a film. Film gamma is given by the slope of the line.

The resolving power of a film is defined by the width of the narrowest pair of lines on a test target, one black and one white, that it can register separately. Resolving power is approximately related to resolution on the ground by the formula

$$g = \frac{H.d}{F} \, 10^3,$$

where g is the ground resolution in metres,
  H the distance from the object to the camera (flying height),
  d the smallest resolvable distance at the image plane in millimetres, and
  F the focal length of the lens.
Film resolution is commonly expressed as the reciprocal of the smallest distance (d) resolvable and expressed as 'line pairs per millimetre'. In this case the formula becomes

$$g = \frac{H.10^3}{F.n},$$

where $n = 1/d$ in line pairs per millimetre. The resolution achieved in practice is always less than the potential resolving power of the film because the resolution of a photograph is the product of the resolving power of both the film and the camera.

Camera lenses may cause loss of resolution owing to various imperfections. These include:

Spherical aberration. Different focal lengths corresponding to different radii in the lens cause light rays following different paths to come to focus at different distances from the principal plane (Fig. 2.4a).

Curvature of the field. Oblique rays come to a focus short of the image plane (Fig. 2.4b).

Coma. Spherical aberration causes oblique rays to form image points that are elongated in one direction.

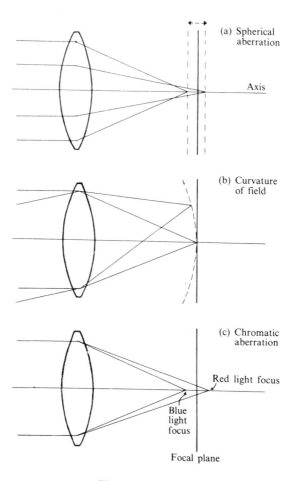

FIG. 2.4. Lens distortions.

Astigmatism. Blurring is caused by the focus in one plane being shorter or longer than that in the plane at right-angles to it (the orthogonal plane).

Chromatic aberration. Light of different wavelengths is differently refracted and brought to a focus at slightly different positions (Fig. 2.4c).

In practice the resolving power of a camera is also affected by the contrast and shape of the target, by the exposure level, and by colour of the light. Under operational conditions it is also affected by blurring of the image induced by camera movement. In aerial photography this is an important consideration, particularly in low-level work. It also imposes a requirement for stabilization in rocket and other aerial platforms.

As indicated above, the grain size of the film affects its resolution: One of the effects of this is to give rise to fluctuations in density that can obscure very fine detail, particularly in areas of low contrast. Exposure and development of the film also affect the final resolution in the photographic product because of the 'spread function' in the reduction of the halide grains. There is also a general, low background density in all negatives resulting from their chemical make-up. This is known as 'fog' and is represented by the horizontal toe of the log e curve of the film (Fig. 2.3).

The best resolution is obtained with targets that are in high contrast to their backgrounds. When film resolutions are quoted they should always be qualified by a contrast figure given in terms of film density; e.g. '150 lines/mm at 1·6:1 contrast'. The contrast figures most commonly used are 1·6:1, 6·3:1, and 1000:1.

Table 2.1 compares the resolution and relative speeds of some commonly used aerial survey films. Note that the Zeiss Pleogon and Wild Universal Aviogon lenses of 152 mm (6 in) focal length used in many aerial survey cameras have resolving powers of only about 40 line pairs per millimetre.

Table 2.1. Resolutions and speeds of aerial films

Film type		Line pairs per millimetre at contrast density 6·3:1	Relative speed
Kodak 3404	High definition	550	1
Kodak SO−243	High definition	440	1·2
Kodak SO−190	Special fine grain	180	3·7
Kodak 3400	Panatomic X	150	9·0
Kodak 3401	Plus X	100	33·0
Kodak 5425	Super XX	75	41·0

In the field the objects to be photographed are illuminated by light from two main sources: from direct sunlight and by diffuse light from the sky and reflectance from clouds. The intensity and spectral content of the light reflected from the target is also affected by atmospheric absorption between the target and the camera. This, and the diffuse haze caused by atmospheric dispersion, can reduce the image quality considerably. However, the atmospheric dispersion due to Rayleigh scattering is greatest at the blue end of the spectrum, so that by inter-posing optical filters which cut out part of the blue light the effect of haze on the photograph can be reduced considerably. The wavelengths and quantities of light absorbed by optical filters, which may be of glass or gelatine, can be plotted as spectral transmission curves (Fig. 2.5). On these the slope of the curve indicates the sharpness of the spectral cut-off.

Sharp-cut filters are required for precise selection of spectral bands. For practical purposes a 10 per cent transmission is regarded as the cut-off point.

Filters are also used with 'normal' film types, sensitive to the visible spectrum, to balance the proportions of recorded light to correspond more closely to the sensitivity of the eye. Introducing filters reduces the amount of light received at the image plane. The exposure is increased to compensate for this, either by increasing the exposure time by adjusting the shutter speed or by increasing the lens aperture.

The aperture, or area of open shutter, is conventionally specified by an 'f' number, which is the focal length of the lens divided by the diameter of the aperture. The illumination is proportional to the area of the aperture; and the diameter of the aperture is proportional to its square root. As it is convenient to measure increase or decrease of illumination by factors of two, this means that the 'stops' or shutter designations increase in powers of the square root of two. Thus the conventional sequence of stops is: 2, 2·8, 4, 5·6, 8, 11, 16, 22.

Four main types of film are used in aerial surveys: panchromatic, infrared black and white, true colour, and infrared false colour.

Panchromatic film is the conventional black and white type, sensitive in the visible range from 360 to 720 nm. The long-wave (red) sensitivity is somewhat extended in aerofilms to compensate for the use of more severe blue-cut filters than is normal with films for use on the ground. It is standard practice to use a yellow (i.e., minus blue) filter with a cut-off at 470 nm for aerial work.

Infrared black and white films have sensitivities extending further into the long-wave region, typically from 360 to 900 nm, which takes in a large part of the 'reflective infrared window'. To obtain the maximum benefit from this characteristic the minus-blue filter should be stronger than for panchromatic emulsions. This imparts excellent haze-

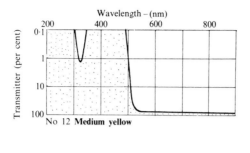

No 12 **Medium yellow**

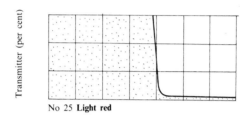

No 25 **Light red**

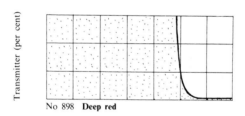

No 898 **Deep red**

FIG. 2.5. Spectral transmission curves for filters (Kodak Wratten series). No. 12, medium yellow, is used with panchromatic and colour infrared films to reduce atmospheric scattering effects; No. 25, light red, has a much greater haze-cutting effect and is used with panchromatic and black and white infrared films; No. 89B, deep red, is a visually opaque filter used exclusively with black and white infrared film.

cutting characteristics by eliminating much of the obscuring effect of short-wave light scattered by the atmosphere. It has, however, the disadvantage that areas in shadow, which are illuminated largely by reflected and scattered light of short wavelength, appear very dark on the exposures and detail in them is poorly resolved. Most available types of infrared black and white film have lower resolution than most panchromatic films: 55 line pairs/mm at 6·3:1 contrast is typical.

Water absorbs infrared radiation strongly. Bodies of open water and moist soils are therefore more strongly expressed on infrared than on panchromatic film, appearing very dark in the positive. On the other hand, live vegetation has a very high infrared reflectance and appears bright in comparison with its normally dark tones on panchromatic film.

True-colour films have three-layered emulsions, the halide grains in each layer being sensitized by 'coupling' with a complementary dye to register one of the primary visible wavelengths (blue, green, or red). On development of the exposed film a silver image is formed in each layer in the same way as in black and white processing, but it is combined with a coloured dye image. Further processing removes the silver and leaves the three dye images, which contain all the information required to produce the photographic scene in colour. In a true-colour emulsion the blue sensitive layer is in the outer position and the red-sensitive layer is closest to the film base. An anti-halation layer below the emulsion prevents the back-scatter of light from the film base material. A yellow or orange masking filter in the film below the blue layer improves colour fidelity by reducing the amount of short-wavelength light passing the blue layer (Fig. 2.6). In some of the colour film used to photograph the Earth from the *Gemini* and *Apollo* space satellites

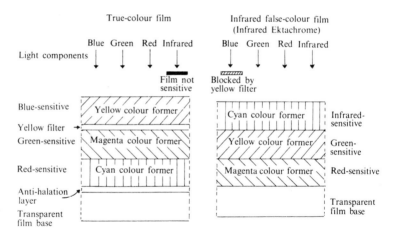

FIG. 2.6. Colour film construction. With colour negative films, development produces a negative image in the respective layers. In colour 'reversal' films the images formed are positive. Infrared Ektachrome is a reversal film. The yellow filter in the true-colour film prevents blue light exposing the green- and red-sensitive layers. For Infrared Ektachrome film a yellow filter is used in front of the camera lens. An anti-halation layer is used in some films to prevent light reflecting back from the film base.

the normal order of colour sensitized layers was reversed, the red sensitive layer being uppermost. This was done to record the maximum amount of long-wavelength light in ·order to reduce the modifying effects of the full column of the atmosphere through which the light had to pass. The results from these films were not, of course, strictly true-colour.

There are two kinds of true-colour film.

In *colour negative films* the image is initially formed in its complementary colours, that is in yellow, magenta, and cyan, which are complementary to blue, green, and red. To produce a true-colour positive print, the image has to be transferred to another emulsion.

Colour reversal films, on the other hand, form a positive image from the silver halide grains remaining after the initial developing process to produce a true-colour image in the form of a transparency.

True-colour films have approximately the same range of spectral sensitivity as panchromatic films. They are, however, somewhat slower than normal black and white emulsions and they require good lighting conditions when used in aerial survey. This can be a considerable problem in temperate climates with variable weather. In the United Kingdom, for instance, it is very difficult to obtain satisfactory colour photographs in the winter months.

False-colour film is a multi-layered emulsion which incorporates a near-infrared sensitive layer. Since the radiation recorded in this waveband is invisible, it must be assigned a colour to be made visible. This prevents the normal true-colour assignment of the three primary colours to their respective wavebands. *Infrared Ektachrome* was originally developed from camouflage detection films. In this, near-infrared radiation is registered in red, red light in green, and green light in blue. Blue light is prevented from exposing the film by stronger yellow filters than are used with true-colour film. These are normally Wratten 12 or 14 filters with a cut-off at about 550 nm.

False-colour film thus renders familiar scenes in unusual colours. This is particularly evident in vegetated areas (for live vegetation has high reflectance in the near-infrared), so that false-colour film shows vegetation in various shades of red rather than the natural greens (Pl. 1).

Since Infrared Ektachrome is a colour reversal film, the colours representing the three wavebands are developed directly on the original film without being re-exposed. If positive prints are needed an intermediate negative has to be made from what is a positive image film. (Mott 1966; Marshall 1968; Heller 1970).

The Russian Spectrozonal film is another type of multi-layered false-colour infrared emulsion, in which dyes are introduced into the separately sensitized layers during the development process. The resulting effects are quite different from those obtained with Infrared Ektachrome; for instance, conifers appear green, while broad-leaved deciduous trees are yellow or orange.

As well as by layered emulsions, colour information can also be recorded by *multi-spectral photography* in which a number of separately filtered exposures are simultaneously made of the same scene. Most simply the scene is recorded through blue, green, and red filters, to give a separate exposure for each colour. Normally a fourth simultaneous exposure is also made with infrared-sensitive film filtered to cut out all visible light. The separate exposures are recorded in black and white (Pl. 2) and may be recombined in one print during processing. Alternatively, special viewers project transparencies through appropriate filters and create true or false-colour simulations in a number of permutations from the four-channel pictures.

Multi-spectral photography may be obtained with a single four-lens camera with a single focal plane shutter to record all four channels of information on one film; or by a pack of four separate cameras with synchronized shutters, which gives separate waveband images on four rolls of film. A refined approach to multi-spectral photography is to record the spectral characteristics of the different elements of the target area using a spectro-radiometer. These data are then used to select or design special-purpose filters to make up the best possible combination for maximum discrimination of the features of interest.

Photography is cheapest with panchromatic or infrared black and white film. Colour and false-colour films are considerably more expensive, both to purchase and to process. Black and white films, moreover, generally have higher resolutions than the colour types, which are slower and more difficult to use for aerial photography owing to their limited exposure latitudes. There have, however, been marked improvements in the resolutions, sensitivities, and colour fidelity of colour and false-colour films in recent years and some materials with very high resolutions are now becoming available. For some purposes multi-spectral photography may be cheaper and more versatile in obtaining colour information than colour photography, since only black and white processing is involved.

The most commonly used photographic product is the familiar positive paper print. These can be contact prints, at the same scale as the negatives (acquisition scale), or they can be enlarged to a more convenient working scale. It is not normal practice to enlarge standard 9-in aerial survey photographs, though photographs in smaller formats (70 mm or 5 in) may be considered too small for convenient handling. Enlargements are always more expensive than contact prints, and secondary processing tends to reduce the resolution of the original.

Table 2.2 Characteristics of main film types

Panchromatic (black and white)
Good definition and contrast
Wide exposure latitude
Inexpensive

Infrared black and white
High (possibly excessive) contrast with loss of shadow detail
Difficult to determine correct exposure
Gives particular emphasis to vegetation and water
Inexpensive

True-colour
Good contrast and tonal range but less good definition than panchromatic film
Wide exposure latitude with negative film types
Expensive

Infrared false-colour
Lower resolution
Vegetation, water, and moist soil emphasized
Film slow and exposure difficult to determine
Purchase and reproduction both expensive

An alternative to the opaque paper print is the positive transparency. This may be the initial film product, as is the case with true-colour reversal films or Infrared Ektachrome, or it can be produced directly from a negative. Film transparencies are more difficult to handle and work with than paper prints but in some instances colours can be more true and images sharper.

It is bad practice to use the original films for interpretation, for they are unique records and are easily damaged, but the cost of colour processing and the possibility of losing information in processing may make this necessary.

Individual prints can be joined together to make a *photo-mosaic*. In its simplest form this is merely a 'lay-down' of untrimmed overlapping prints all at the same scale. Mosaics of better quality are made by trimming and butt-joining the photographs in the montage. For this purpose all the prints making up the mosaic need to be printed evenly and consistently in order to eliminate as far as possible the juxtaposition of tones that do not match from one print to the next. Individual prints may be rectified for scale and tilt distortions to give better geometric accuracy to the mosaic, and base-map or ground control can be introduced during the laying-down of the mosaic to improve its planimetric accuracy. According to the degree of rectification applied in the assembly the product may be referred to as simply a 'stick-down', a 'semi-controlled mosaic', or a 'controlled mosaic'. The quality of a mosaic greatly depends upon the quality of the photography used, but very high standards of accuracy and appearance are possible. Indeed, photo-mosaics are often used as map substitutes, for they can be re-photographed and reproduced relatively cheaply. If selected cartographic detail (say roads and urban limits) is overprinted with annotations, the result is commonly termed a *photo-map*.

The ortho-photograph is a more recent development. Orthophoto-graphs are made from very highly rectified photographs and have the planimetric accuracy of detailed, large-scale maps. Unlike normal aerial photographs the correction is consistent across each entire frame, so that ortho-photo maps are more accurate than mosaics. A number of optical processing machines have been developed to produce photo-negatives corrected in this way. Their basic principle is that small elements of the original negatives of a stereo-overlap are rectified separately and then fitted together to give the final product. The Zeiss— Jena process, for instance, rectifies the image in parallel strips using direct optical correction, whereas the Hoborough—Gestalt process applies the rectification to hexagonal sections using video techniques. There are other methods under development employing parallel print scanning with computer rectification. Needless to say, because of the complexity of the equipments involved, orthophotographs are expensive compared to conventional photography. Their main role is likely to be in engineering and land registry work, though, if available, they could serve as excellent base maps for investigations into soils and other natural resources.

3. Aerial photography

The aerial photographs normally used for soil surveys have in most cases been obtained originally for photogrammetric mapping. Even when this is not so, most of them have been obtained with mapping cameras and are of equivalent standard, with full stereoscopic overlap between successive prints.

Photogrammetry, which is used to produce a large proportion of the topographic mapping requirements in the world today, is based on stereoscopic principles. In natural stereoscopic vision we perceive depth or distance by the relative displacement of objects seen separately by our two eyes. Photogrammetric techniques similarly create an optical model of the terrain by fusing two overlapping but displaced images of it.

Provided that both the observer's eyes are of equal strength, a stereoscopic or three-dimensional model of the terrain can easily be obtained from two overlapping photographs by means of a simple pocket stereoscope consisting of two lenses in a frame. One photograph is viewed separately by each eyepiece. The two photographs are moved relative to each other until their images, seen separately, fuse; the scene then appears in three dimensions. (Some people are able to fuse images without using a stereoscope simply by juxtaposing two prints and viewing one with each eye.)

The stereoscopic plotters used for topographic mapping employ this same principle with great precision to provide accurate measurements and plotting of detail. In these instruments pairs of photographs, rectified to a required scale and to remove distortions caused by tilting of the camera at the time of exposure, are mounted either as glass-based transparencies (diapositives) or rigidly between glass. The operator sees the photo-pair through a binocular eyepiece with a limited field of view and moves the carriage on which it is set up to cover the rest of the overlap. A spot in the centre of the viewing area is the reference point for tracing detail. As this is moved across the overlap its position is tracked by arms or 'space rods'. Adjustable linkages to a plotting table enable detail to be traced at various scales. Further, by using the principle of parallax displacement, separate spots on the left and right viewing optics can be fused and manipulated to track across the stereo-model at a predetermined height to draw contours.

Special high-performance cameras and systematic precision flying are necessary to obtain photographs to the stringent requirements of photogrammetry. Most aerial survey cameras use film 9 in (240 mm) wide and have a focal length of 6 in (152 mm). Cameras with very wide-angle lenses with focal lengths of 3·5 in (88 mm) are sometimes used for small-scale coverage of very large areas.

The scale of aerial photography depends upon the flying height of the aircraft above the ground and the focal length of the camera:

$$\text{Photo-scale} = \frac{\text{Flying height}}{\text{Focal length}}$$

Thus, with a lens of 6 in focal length, the 'acquisition' or 'contact' scale of the negatives will be double the flying height in feet. If the aircraft flies at 5000 ft the scale of the contact prints will be 1 : 10 000. With an 88-mm super-wide-angle lens at the same height the acquisition scale will be 1 : 7240. Table 3.1 lists the photo-scales for various flying heights.

Even with an aircraft flying a flat and level course, the scale of photography will vary across each print, or frame, with variation in height above sea-level of the ground below. In mountainous areas this may give rise to large discrepancies in scale on one photograph.

Aerial survey cameras are by no means light-weight instruments. A typical camera with a fully loaded magazine is about 60 × 60 × 60 cm overall and weighs over 100 kg. It requires a specially adapted aircraft with a camera mount and a hole cut in the floor. Twin-engined aircraft of medium size are generally preferred. Air survey companies operating in the United Kingdom use aircraft such as the De Havilland Dove, the Britten–Norman Islander, and the Beechcraft Queenaire as well as some single-engined types. Aircraft with greater range like the twin-engined Dakota DC 3 are better for covering very large areas. The French Institut Géographique National (IGN) use four-engined Boeing B 17 Flying Fortresses of Second World War vintage for some work.

Light twin-jet or turbo-prop aircraft are being used increasingly for surveys of very large areas of over 10 000 square kilometers, which usually call for high-altitude, small-scale photography. Though they have relatively short endurance such aircraft have high operating speeds and a fast rate of climb which enables them to reach their operating altitude quickly. Aircraft such as the Gates Lear Jet and the Rockwell Jet Commander or the turbo-prop Air Commander are used commercially. The Royal Air Force employs twin-jet Canberras and four-jet Victors. Perhaps the most celebrated of all photographic aircraft is the Lockheed U 2, a twin-jet military reconnaissance aircraft with a high performance that can operate at very high altitudes. The United States National Aeronautics and Space Agency uses the U 2 for civil experimental work but they are not yet in commercial operation. Jet aircraft

Table 3.1. Altitude, scale, and coverage for standard aerial survey cameras

| Scale of photography | Flying height | | Width of ground cover strip (km) | Area covered by one 9-in print (km²) |
	Focal length: 6 in (152 mm)	3·5 in (88 mm)		
1:5000	760 m (2500 ft)	440 m (1458 ft)	1·14	1·31
1:10 000	1520 m (5000 ft)	880 m (2917 ft)	2·29	5·23
1:20 000	3040 m (10 000 ft)	1760 m (5833 ft)	4·57	20·90
1:30 000	4560 m (15 000 ft)	2640 m (8750 ft)	6·86	47·03
1:40 000	6080 m (20 000 ft)	3520 m (11 667 ft)	9·14	83·61
1:50 000	7600 m (25 000 ft)	4400 m (14 583 ft)	11·43	130·64
1:80 000	12 160 m (40 000 ft)	7040 m (23 334 ft)	18·29	334·45
1:100 000	15 200 m (50 000 ft)	8800 m (29 167 ft)	22·86	522·58

are too fast for low-altitude survey as opposed to military photo-reconnaissance work.

For missions of relatively short duration at low altitude the normal air crew will consist of a pilot, navigator, and camera operator. A second pilot will be carried for longer missions. Aircraft used for high-altitude aerial surveys have to be pressurized and this creates problems with cameras. If mounted outside the pressurized compartments they have to be operated by remote control and the lenses are prone to condensation. If the cameras are mounted inside the pressurized compartment a plate or 'flat' of optical glass must be placed across the camera hole. Glass flats that will not affect the quality of the photography by introducing distortions in one form or another are extremely expensive.

Block coverage is obtained by flying at a constant altitude (to maintain an even scale of photo-cover) in regularly spaced parallel flight lines. The camera shutter is operated electrically and its timing is adjusted to obtain the required degree of overlap at the ground speed of the aircraft. For full stereoscopic coverage at least 60 per cent overlap along the track of the aircraft is necessary; as much as 90 per cent overlap may sometimes be required. Lateral overlap between parallel flight lines is normally between 10 and 15 per cent. This ensures that no gaps are left by flight-line separation and allows the transfer of planimetric ground control from the prints of one run to those of the adjacent run (Fig. 3.1).

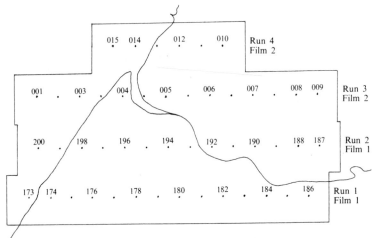

FIG. 3.1. Aerial photo flight plot.

The ground coverage, or field of view of a single frame, is determined by the focal length of the camera and the dimensions of the film. A camera of 6 in focal length with a film 9 in wide flown at 5000 ft will cover a ground strip 7500 ft wide. In practice, because of side winds, successive overlapping prints are often offset to some degree and do not have exact lateral coincidence along the flight track. Even if the pilot can maintain an exact heading in a crosswind, the aircraft will move away laterally from the intended flight track, and if he alters the heading the aircraft will fly crabwise. This crabbing effect can be cancelled out to some extent by rotating the camera in its mounting (Fig. 3.2). In most aerial survey operations the camera operator controls the camera rotation manually, using a drift sight to keep the camera axis parallel to the line of flight.

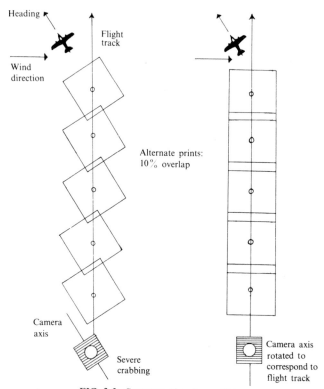

FIG. 3.2. Compensation for crabbing.

For photogrammetric mapping, ground control is obtained by surveying-in ground targets specially placed at easily recognizable points. This improves the accuracy of the mapping and makes it possible to compensate for such factors as lens distortion, lack of flatness in the film, and the curvature of the earth. The number of ground control points needed on each print is six: four in each 60 per cent forward overlap area and two in each lateral overlap. If the adjacent runs are staggered, four more points may be necessary to carry forward the lateral tie-in. Where the photography is required only for surveys of soils or other natural resources, it is not normally necessary to have ground control.

Because most films used for aerial survey are fine grained, in order to obtain the best resolution, they are relatively slow. To obtain satisfactory exposures, shutter speeds and apertures are therefore critical. These are, however, affected by a number of factors. First, the shutter speed must be fast enough to prevent the image becoming blurred by the motion of the aircraft relative to the ground. The faster and lower the aircraft flies, the more apt this is to happen. Shutter speeds of down to 1/1000 s are usual for panchromatic films, but even so, photographs cannot normally be taken below 1200 ft above the ground with most survey aircraft. Blurring can also occur if turbulence causes the aircraft to pitch, roll, or yaw at the moment of exposure. Gyro-stabilized camera mountings are available to counter some of these effects but they are rarely used in routine operations because the displacement of the image caused by the instability of the aircraft can be compensated for in the stereo-plotters if the photography is for photogrammetric mapping. Distortions caused by camera motion, and even film resolution, are generally not limiting factors when the photography is to be used for purely interpretive purposes, as in soil mapping.

Time of day or year, weather, and the state of the atmosphere affect the quality of aerial photography and are obvious considerations in aerial survey operations. Even a slight haze can reduce the quality of the exposures drastically. Photography from below an overcast sky requires an increase in exposure time, which may be ruled out by the danger of introducing blurring through image motion. In high and low latitudes the level of illumination depends very much on the elevation of the sun. Sun angles must be greater than 30° for routine survey operations in the United Kingdom. Low-altitude photography using colour or false-colour films, which are particularly slow compared with panchromatic films, is rarely practicable in Great Britain between the end of September and the beginning of April, except possibly during one or two hours either side of noon in exceptionally clear conditions.

Edge shading or 'vignetting', which is due to lack of transmission at the edges of the lens, causes loss of image quality, particularly with colour films. This has to be overcome by the use of anti-vignetting filters which have an equal and opposite effect on the transmission. The use of these and any other filters reduces the quantity of light received and they may have to be compensated for by a slower shutter setting or wider aperture.

Once obtained, aerial photographs require high standards of processing to preserve their quality. Properly controlled developing and printing are particularly important for colour and false-colour films if consistent results are to be obtained along a complete roll of film. Hand processing is still used extensively and can give excellent results, but there is an increasing trend towards machine processing. Electronic printers scan the film and balance the densities using a modulated light source. This produces prints that are even in tone across the frame and are also in tone-matched sets. To obtain the strict control of temperature and timing needed for colour processing, automatic processors like the Versamat are used. Versamat processors and electronic printers are extremely expensive but the quality of the product justifies their use, particularly when a large investment has been made in the survey in the first place.

Aerial photography is readily commissioned from aerial survey companies or the equivalent government organizations (see Appendix 1). Such is the flexibility of aircraft operations that this applies equally whether very small or very large surveys are required. The cost of aerial photography varies considerably, however, according to the scale required and the size and location of the area to be surveyed. Large-scale coverage of small areas is much more costly for a given area than small-scale coverage of large areas, for with these the mobilization cost of the aircraft is not such a proportionately large item. In general, panchromatic photography of small areas in the United Kingdom, to full photogrammetric specification and at medium scales (between 1:5000 and 1:10 000), is six to eight times more expensive than equivalent cover of very large areas (25 000 km^2 or more) overseas at scales in the region of 1:50 000.

Because of the higher cost of the films and their processing, colour and false-colour photography are more expensive than panchromatic cover of equivalent scale. This is particularly evident in the prices for the coverage of small areas. For these the rate for colour photography is about 15 to 25 per cent higher than for panchromatic cover. This differential is less for larger areas, and colour coverage may be only a few per cent more expensive than panchromatic for surveys of very large areas.

If the weather is fine and the aircraft is stationed close to the survey area, most flying projects can be completed in a matter of days. In the United Kingdom, however, one may have to wait for months until good flying weather coincides with the availability of an aircraft. The alternative to commissioning one's own coverage is to purchase copies of existing photographs. There is now aerial coverage of some sort for most parts of the world. It may not be at the ideal scale, it may be considerably out of date, and it may have been taken at the wrong time of the year for the application required, but it will be very much cheaper than specially commissioned coverage.

The whole of the United Kingdom has been flown at various dates during the past 20 years — some areas many times. Prints can be purchased from the libraries of the air survey companies and from the national collections of the Department of the Environment in London and the Scottish Development Department, Edinburgh (see Appendix 2). Both these last centres also maintain a complete register of available cover. Some overseas cover may be found in the libraries of the air survey companies as well as in national libraries. A small royalty may be payable to the original client in some circumstances. This is notably the case with block coverage of the United Kingdom flown for county authorities.

4. Aerial photography in soil survey

In typical soil surveys aerial photographs are interpreted to define the boundaries of soil units, or the land-forms corresponding to them, and the land classification boundaries that are derived from both. Interpretation of the photographs is only one of the survey operations and cannot be a substitute for field-work. It may, however, enable work in the field to be reduced, better planned, or otherwise rendered more effective.

Aerial photographs are not used to actually identify soil types, but rather to locate changes in the land-surface patterns that may be relatable to differing soil properties. In ideal circumstances photographs record differences in reflectance from vegetation or the soil surface that can be correlated directly with the boundaries of mappable soil units. Having a permanent record on a photograph that approximates in geometry to a map can be an enormous advantage in soil mapping. However, the pattern of tonal changes does not tell the interpreter directly what the soils are in detail; this requires examination on the ground.

As well as for interpreting soil boundaries and planning field-work, aerial photographs are used for field navigation and plotting. When employed in this manner to locate sites to be investigated by soil pits or borings, they are used purely as map substitutes. In showing every small ground feature photographs are preferable to the best maps. The slight discrepancies in scale are unimportant when positions are located by direct visual reference to ground features and distance measurement is by pacing or by odometer readings on a vehicle. Precise contour information is equally unnecessary, for a stereo-view can be obtained easily in the field with a pair of overlapping prints and a pocket stereoscope. Contact prints are normally used for field-work. Because of their cost and the risk of damage to them, colour prints are not normally used in the field, but sets of black and white prints can be made from colour negatives for this purpose.

If the area to be traversed is extensive and the scale of the photography is also large, a considerable number of prints may be necessary for a day's work. In such cases field mosaics are preferable. These would normally be linen-backed and foldable. A stereo-view can still be obtained from a mosaic with alternate loose prints at the same scale.

This is not, however, essential for navigation in many areas of even terrain. Even in mountainous areas experienced surveyors can work satisfactorily with single-sheet mosaics alone. Mosaics are, of course, more expensive than photo-prints but paper copies of them can be produced very cheaply from screen-printed negatives. Though the clarity and resolution of these is inferior, they are usually adequate for field use and the compilation of base detail.

The use of aerial photographs as base maps is of most value in areas where the basic topographic mapping has never been carried out, where the available maps are substantially out of date, or where large tracts of land have very few features of the kind that are normally plotted on conventional maps. For example, even drainage patterns are lacking over very large areas of the Clay Plains of the Sudan and parts of the Mesopotamian Plain. In areas such as these, up-to-date photographs showing every bush and hummock in detail are essential for accurate site location and navigation. Especially with reconnaissance surveys, not the least valuable of the uses of photography is to locate the most feasible access routes into difficult terrain. Sometimes the ground conditions prevent the planned investigations from being carried out at the time intended; flooding by surplus irrigation water is, for instance, all too familiar. Under such circumstances the aerial photographs enable rapid decisions to be taken on alternative work plans.

Panchromatic aerial photographs give information on relief, surface texture and tonal patterns, and the relationships of objects and surfaces to each other. Colour and false-colour photography give the relative spectral values of the different components of the scene.

Texture is the pattern of variation of tones over the surface of an area. Tonal patterns represent the light-reflectance characteristics of the various surfaces, but on a micro-scale compared to the broader light and shade patterns due to relief. The appearances or expression of surfaces or objects by which we recognize them are referred to in current terminology as 'signatures'. With soils, as with most other things, they may or may not be unique. In the simplest case a single tone or shade of grey may be adequate to identify a feature. Colour, which is more likely to give a unique signature, is simply a combination of the tones of the three primary colours. Except for single-point targets, both the tone and the texture constitute the signature. This concept of signature is discussed further in Chapter 8.

Though the tones and fine textures on photographs are important, the interpreter will usually obtain more valuable information from the sizes and shapes of objects and their relationships to each other; that is, the broad spatial information in the scene. This is clearly demonstrated when we consider that we are readily able to identify familiar objects on photographs even when they are rendered in the wrong colours, are under- or over-exposed, or are printed in negative form. The fact that we can cope even with severe spatial distortions illustrates the remarkable capacity of the human brain for unscrambling picture data.

Though it is the identification of discrete objects like hills, roads, and rivers that is of paramount importance in navigation, soil mapping requires the location of the boundaries between soil units. In practice, boundaries can be distinguished only if they correspond to discontinuities in tone, colour, pattern, or slope. Even where they are recognizable, such discontinuities are by no means consistent, nor does their strength of expression depend on the category of classification: some soil phase boundaries may well be more obvious than those of some great soil groups. If photo-patterns are positively correlated with the boundaries of soil units, mapping can be far more rapid and accurate than it would be if they were traced on foot over long distances. The value of photo-interpretation depends as much on the conditions in a particular area as on the requirements of the survey. The closer the soil units correspond to land-forms, or to other visible features, the greater the value of aerial photography for mapping. If the soil differences are predominantly chemical or textural, or are based on subsurface horizons or soil depth, and if these factors have no surface expression, the boundaries cannot be seen on photographs and must be interpolated between sample sites.

The photo-patterns by which soil units can be identified may express a number of different surface factors. Changes in vegetation cover are usually well expressed on photographs and often indicate mappable soil boundaries. In undisturbed conditions changes may be due to different plant associations growing on different soil-types, or to slight variations in species composition or plant vigour resulting from different moisture conditions in adjacent soils. Vegetation indications of soil differences are less good where there is intensive cultivation or in the dense vegetation cover typical of humid climates. In general photo-interpretation tends to be easier in arid climates where the vegetation modifies the appearance of the ground to a lesser extent.

Drainage and erosion patterns, or changes of slope, which indicate the nature of drift deposits or underlying geology, and the land-forms which result from them, also reveal the distribution of soils. Micro-relief often distinguishes mapping units at soil-phase level.

On bare ground, such as ploughed or eroded land, soil colours may give direct indications of the soil type, but they are of rather less value than might be expected. Only the colours of the surface horizons are recorded and these may not be typical of the whole soil profile. In arid areas devoid of vegetation the very thin light-coloured rain crusts tend to obscure the colours even of the surface horizons.

Photo-interpretation

A stereoscope must always be used for the systematic photo-interpretation of a project area if the maximum possible amount of information is to be obtained. Working with batches of loose prints of standard aerial survey cover, each 60 per cent overlap of adjacent prints in a run should be interpreted in turn. Once a run has been completed the interpretation can be linked to that of the next run across the 10 per cent lateral overlap by transferring the details at the edges.

Large mirror-stereoscopes with variable magnification are generally preferred for systematic work of this kind. They have a much larger field of view than pocket instruments, and with them one can view comfortably almost the entire overlap of a standard pair of 9-in photographs. Some instruments have rolling mounts to avoid resetting the photo-pair each time the stereoscope is moved. An alternative arrangement is to mount the photographs on a board that can be slid about between the legs of the instrument. An entire mosaic sheet can be interpreted with a minimum of effort by setting it up under the larger 'roller-coaster' or scanning type of instrument and using individual loose prints to provide a succession of stereo-models.

Photo-interpretation proceeds by determining which of the features on the photograph are significantly related to soil boundaries and to the characteristics of the soils themselves. The interpreter's conclusions will be based on his own experience and the knowledge of local geology, land-forms, and ecology he had derived from published work and from his own field-work.

The first aim is to identify the broad general relationships of all factors contributing to the landscape and thence to distinguish these definable units of land with common features. Conclusions can then be drawn regarding the nature of the soils of the individual units. For instance, alluvial soils are generally deep and soils on units of hard rock showing features such as jointing or karst can be expected to be shallow. This information, on the soil characteristics, supported by field survey, enables the interpreted units to be used as a framework for a soil map, the soils of each unit being defined in terms of the soil classification adopted. This will often entail the grouping or the subdivision of photo-interpreted units. If, however, this is carried too far it will discount the value of the photographs. It is far better to 'map what is mappable' than to try to impose an arbitrary system of classification not directly related to what can be seen on the photographs or in the field. The interpreted units should, where necessary, modify any pre-conceived classification.

The compilation may be drawn on overlays of alternate prints of the stereo-pairs or on an overlay to a photo-mosaic. In the first case, the information on the individual overlays must eventually be transferred to a single map sheet. Photographic distortions often prevent accurate transfer by simple tracing, but compilation instruments such as the 'Sketchmaster' or a 'Zoom Transferoscope' can be used. By projection or angled viewing these instruments remove photographic distortions caused by tilt and the changes in scale caused by changes in altitude. Some enable compilations to be made at scales different from those of the photographs. Compilation of maps on controlled or semi-controlled mosaics is commonly used in developing countries and other areas where adequate base maps do not exist.

Good photo-interpretation requires a high degree of concentration and reasonable working conditions. Though pocket stereoscopes can be used in the field it is more convenient and effective to carry out detailed photo-mapping in the office or base-camp, restricting field interpretation to the examination of local problems. In the field the main use of the photographs is for navigation and plotting.

Ideally the use of aerial photographs in soil mapping projects should proceed in four stages. The first stage is planning. This entails a preliminary examination of the complete photo-cover of the survey area and its environs, with or without a stereoscope, in conjunction with any maps that are available. The purpose of this is to get a general 'feel' for the area, in terms of its geography, topography, geology, geomorphology, and vegetation cover.

It is usual to draw up preliminary sketch-maps to show the principal topographic features at this stage, and to decide the layout of the final soil map sheets. The scales and sizes of the final map sheets should also be selected if these have not already been determined. It is often an advantage to compile a provisional geomorphological interpretation on overlays to mosaic sheets.

From this general familiarization with the area a provisional plan of campaign for the survey is drawn up in the following terms:

Access into and within the area;

Selection of traverses and/or layout of a survey grid;

Selection of sampling areas and of priority areas for more detailed investigations.

Actual conditions in the field may force modifications, however. Access, for instance, may not be what it seems on the photographs, especially if old photographs are being used.

For reconnaissance surveys the selection of representative sample areas for detailed investigations should be reasonably comprehensive at this stage, since the selection will be based on geomorphological criteria and will not normally be changed as a result of further examination.

At this stage of the work it is best to mark the photographs or mosaics as little as possible, using easily erasable or changeable methods to allow for the inevitable alterations. Transparent overlays may be used but it may be more convenient to use wax or cellulose-based pencils directly on the photographs. These can be erased without damaging the surface.

At the end of the planning work the surveyor should have a good impression of the geography of the area, preliminary sketch-maps, a plan of execution for the survey, and a set of field prints marked up for investigation. The next stage is field-work.

As indicated above, the photo-cover in the field will be used mainly for navigation from site to site or along traverses and will be annotated with the reference codes for the completed investigation sites. In addition it may be necessary to mark in local changes in geographical or land-use detail: for instance, new roads or farms, particularly if the photography is out of date.

Whether loose prints or mosaics are used, to save damage to the field materials by excessive handling it is advisable to take out only enough cover for one day's work. In unfamiliar areas it is, however, a good idea to take enough material for an alternative sortie. Time can then be saved if the planned access proves not to be as expected.

During the course of the work continuous reference to the photographs progressively clarifies the relationships between soil, geomorphology, vegetation, and land-use, building up the experience which will be used in the compilation of the final map.

After the field-work post-field plotting is used to keep a check on the progress of the survey. It also provides an alternative record of the work should anything happen to the working materials. This procedure is particularly important for projects on which a team of surveyors is employed, and it enables the project leader to impose a consistent approach. Usually the locations and codings of the investigation sites that have been visited, and the mapped soil boundaries, are transferred to an office copy of the mosaic or field map on the completion of each day's work. Any provisional classifications of soil units or individual profiles, made before the results of laboratory analyses are available, may be added.

Periodically a provisional soil map is compiled for the area so far completed. This can give early indications of problems of interpretation arising in the mapping so that additional investigations can be undertaken to clarify them before the field-work is completed. This is particularly important in remote areas to which it may be impractical to return later. The provisional classification provides a list of mapping units which can be amended as the survey proceeds. Mapping units must be mappable, irrespective of the formal soil classification. What units are actually mappable depends not only on their distribution and size, but also on the density of the investigation sites and the scale of the compilation.

The final soil map is compiled by detailed photo-interpretation after the completion of the field-work for the entire project or an individual block or map sheet. This requires a synthesis of all the soil, land-form, and vegetation information collected during the survey, together with the results of any chemical analyses.

Ideally, the field prints, the working compilation material, and the final mapping will all be at the same scale. This should be appropriate to the detail of mapping, i.e. to reconnaissance, semi-detailed, or detailed surveys. The detail, or level, of a survey is usually defined by the density on the ground of the investigation sites to be described and sampled, or by the spacing of traverses and the frequency of sampling along them. Though there are differences in detail there is general agreement among various authorities on specifications and terminology for levels of survey. Examples of specifications for the various levels of survey are given in Table 4.1. In practice it is unlikely that aerial photography will be at contact scales of less than 1 : 100 000, and surveys at reconnaissance and lesser densities will be compiled on reduced mosaics. Satellite photography and other types of imagery, including side-looking radar, may fill the requirement at these scales.

In practice, ideal procedures cannot always be followed. This is particularly true where the specifications for the soil maps are adjusted to take account of the requirements of engineering design or farm layout, or where the photographic cover, taken for other purposes than the task in hand, is supplied independently. In this respect government organizations undertaking soil surveys are usually better able than independent consultants to standardize procedures.

Examples of the use of aerial photography

(1) *Iraq*

A project was carried out between 1962 and 1968 to provide soil and land classification surveys for the development of irrigated agriculture on some 5500 km^2 between Mandali and Badra in the eastern part of Iraq. The area is largely covered by alluvial outwash from a complex of fans along the foothills of the Zagros Ranges which extends into the great plain of the Tigris. The soils are predominantly deep and of medium to fine texture. The chief limitation to agricultural development is the salinity. This varies widely and is related to drainage, texture, and land-use.

Table 4.1. Mapping scales for types of soil survey
(FAO agreed)

Type of survey	Mapping scale	Mapping units
Very detailed	Larger than 1 : 10 000	Phases of soil series, occasionally complexes
Detailed	1 : 10 000 to 1 : 25 000	Phases of soil series, complexes
Semi-detailed	1 : 25 000 to 1 : 100 000	Associations of soil series, physiographic units enclosing soil series
Reconnaissance	1 : 100 000 to 1 : 250 000	Association of great groups, occasional individual great groups; phases of great groups
Exploratory	1 : 250 000 to 1 : 1 000 000	Land units of various kinds, preferably enclosing great soil groups
Synthesis	smaller than 1 : 1 000 000	Great groups and phases of great groups

Adapted from FAO Working Party Report *Land classification in feasibility studies for irrigation development.* LA: Misc./69/5 (1969).

The survey procedures and map scales were determined by the engineering design requirements. The first phase of the project, in 1962, was a reconnaissance soil survey of the whole area, the aim being to identify the range of soils and to delineate the areas of better agricultural potential that would merit further investigation. The field-work in this phase was based on a grid of investigation sites at a density of approximately one site per 10 km^2. Panchromatic prints at a scale of 1 : 20 000 taken in 1957 were available for the northern area and 1 : 30 000 cover taken in 1953 was available for the southern area. The only available topographic maps were compiled in the 1920s.

For this phase of the survey generalized soil and land classification maps were compiled as overlays to 1 : 10 000 mosaics made from the negatives that were used for the sets of field prints. Because of the sparsity of ground investigations the interpretation of soil and land class boundaries depended very heavily on geomorphology, on the pattern of cultivation, and on the distribution of the natural vegetation. These last two factors change quite rapidly, and the photography was sufficiently out of date to introduce errors in some areas.

The second phase of the project, carried out between 1964 and 1966, comprised semi-detailed soil survey of some 250 km^2 in the northern area, identified as better-quality land in the first phase. The density of survey was five sites per square kilometre, but investigations were in fact carried out in two stages. Because of an urgent demand to begin preliminary work on planning the irrigation layout, the area was covered at a density of two sites per square kilometre during the first winter working season and preliminary maps were made available to the engineers. The work was brought up to full specification in the following working season and the final maps were produced for the detailed planning of the irrigation and drainage layout.

New panchromatic 1 : 20 000 scale photography was specially flown in 1963 for this second phase. Soil and land classification maps were compiled as overlays to new sets of rectified mosaics at 1 : 10 000 scale and the final maps were also produced at this scale.

These surveys were part of a project to develop a number of new irrigation schemes and they demonstrate a general lack of adherence to theoretically ideal procedures which is quite typical. The textural complexity of the alluvial soils and the dependence of the land classification on the degree of salinity somewhat limited the contribution of photo-interpretation to mapping. The final map scale was probably too large for the density of investigation sites but the determination of the distribution of the soil and land class types was sufficiently accurate for planning the irrigation layouts.

In practice, although the 1963 photographs were useful in that they recorded the more recent cultural changes in a somewhat changeable area, they were less good for photo-interpretation than the older coverage. The new cover had been electronically printed, and although the tones were even the general levels of contrast were reduced throughout. The older cover, though inconsistent in tones between frames, had much higher contrast levels. In areas such as this, subtle contrasts in ground tones reflecting differing surface conditions and micro-relief are of great value to the interpreter, and this was a decided advantage.

(2) *Republic of Niger*

Work for the Dallol Maouri Development Project was carried out under the direction of the Food and Agriculture Organization (FAO) of the United Nations. Soil and land class mapping were required as basic information for planning regional development in an area of 8000 km^2 extending along the Nigerian border from 11° 30′ N to 14° N. The area was centred on the dry valley system of the Dallol Maouri. (*Dallol* is Hausa for valley.) The country-rock is predominantly sandstone, capped by thick fossil laterite. Most of the area is overlain by a fixed sand sheet dating from one of the interpluvial southern extensions of the Sahara.

The survey specifications required: reconnaissance soil and land classification maps of the entire area; semi-detailed soil and irrigation capability maps of 450 km^2 of representative areas covering the full range of soil types; detailed soil and irrigation suitability maps of three pilot areas for groundwater irrigation, totalling 500 ha. This is a complex specification typical of feasibility studies for mixed regional development.

Panchromatic cover at a scale of 1:50 000 was used as a basis for the reconnaissance mapping. The overall density of investigation sites was one per 10 km^2. The ground was covered in a series of traverses and photo-interpretation was used to interpolate soil boundaries between them. Compilation was on mosaics of the same scale as the field prints and the final map was drawn as a single sheet at a scale of 1:200 000.

The semi-detailed survey used 1:25 000 panchromatic coverage of more limited extent than the 1:50 000 photography. The ground investigations were on a grid pattern to a density of four sites per square kilometre with detailed soil boundaries interpreted from the photographs. Both compilation and map presentation scale were the same as the contact scale of the photography.

The detailed soil surveys were carried out using 1:5000 scale enlargements of selected 1:25 000 frames. The investigation site density for these areas was two per 3 ha. At this detail of mapping the main value of the photographs was as field and base-maps rather than for interpretation. All the photography used was specially obtained for the project during the previous year.

Throughout the area the main soil parent materials are sands of different origins but as the annual rainfall varies from 600 mm in the north to 850 mm in the south, a wide range of zonal soils has developed. To allow correlation with previous mapping in Niger the soil classification used by ORSTOM (Office de la Recherche Scientifique et Technique d'Outre-Mer) was followed. For the reconnaissance survey the mapping units were physiographic, corresponding roughly to land systems, separate colour codings being used to distinguish the different soil parent materials. Within this framework 42 separately mapped soil units were distinguished at association, series, and phase level (Table 4.2).

Both the foregoing examples illustrate the value of aerial photography for large projects, particularly for reconnaissance and semi-detailed mapping at small scales. Pre-project feasibility surveys of the type described are very often based on aerial photography. More detailed work, on the other hand, may not benefit from it so much. The author took 1:15 000 scale panchromatic prints into the field on a 6 in to 1 mile survey in the Middle Thames valley but found little use for them, even where there were clear relationships between the soils and the geomorphology. At this scale, close observation, detailed sampling, and very good base maps reduced the advantages of the photographs.

**Table 4.2. Photo-interpretation for reconnaissance survey in the
Dallol Mauori Project, Republic of Niger**

Main units and other mapped information for a 1 : 50 000-scale compilation

Landform units (letter coded)	Number of soil sub-units within the main units
Plateau (laterite cuirasse)	4
Erg (sand sheet)	11
Glacis (intermediate erosion surface)	11
Dallol (dry valley floor)	12
Alluvial formations	4

Soil parent materials
(colour coded)

Residual drift on laterite
Residual drift on ferruginous surface
Wind-blown sand
Drift derived from local sandstone
Valley-bottom sands (Dallol)
Hydromorphic deposits
Alluvium

Colour, false-colour, and multi-spectral interpretation

Although colour undoubtedly gives more information than panchromatic photography it is not yet widely used. Colour and false-colour aerial photography has been generally available only since the early 1960s. Early aerial colour for instance suffered from 'hot' or bright spots in the centre of each frame and it was not until anti-vignetting filters were introduced that this was largely eliminated (see Chapter 3). Unexposed false-colour films also, like all infrared films, tend to degenerate rapidly in storage unless they are properly refrigerated. There are complicated problems of exposure and of colour balance in processing either of these types of film.

Most soil surveys therefore use panchromatic photography, but colour and false-colour photography are likely to become more widely used as their problems are better understood and workers become more used to them. Usually there is more chance of colour photography being used when the coverage is specially commissioned for the survey. For instance, colour and false-colour have both been specified for some soil surveys by the Land Resources Division of the Ministry of Overseas Development and by the FAO. It is still difficult to judge, however, whether colour or false-colour photography is more cost-effective than panchromatic for soil surveys, though in particular instances there are clearly definite technical advantages.

At present multi-spectral or multi-band photography is rarely a realistic choice for routine soil surveys. Research on cover of this type is being carried out but the few results so far published are inconclusive. A serious difficulty is the inconvenient format: three or four separate 70-mm or 5-in film 'chips' or multi-image prints on one frame are much more difficult to deal with in the field than standard aerial cover. Processing can be complex and costly, and if full use is to be made of the colour additive principles for interpretation, expensive viewers must be used. The quality of the photographs also tends to compare unfavourably with that of cover in standard 9-in format. On the other hand, the advantages may be no greater than those of conventional colour and false-colour photography.

5. Linescanners

Linescanners provide one approach to the problem of producing images from parts of the spectrum that cannot be recorded directly on photographic films. The technique was first developed for mapping with thermal infrared wavelengths greater than 1 μm, but it can also be used for other parts of the spectrum. There are broadly three families of scanners: (1) single- or dual-channel thermal infrared scanners; (2) microwave scanners; (3) multispectral scanners operating in the ultra-violet to near-infrared atmospheric window, which includes the full visible spectrum.

Linescanners, whether single- or multi-channel, collect their information in the object or target plane rather than in the image plane; that is, the information is recorded directly from the target, segment by segment, rather than by 'fixing' a complete image in the focal plane as with a camera. The complete image is not registered in one instant by the use of a shutter, but is built up from a succession of lines scanned beneath the aircraft. The image data are collected by an optical system composed of mirrors or prisms rotating or oscillating at right-angles to the direction of flight (Fig. 5.1). These operate at speeds of between 300 and 3000 cycles per minute.

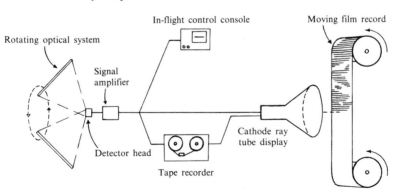

FIG. 5.1. Components of a linescanner.

Like photographic and video cameras, linescanners are passive instruments: they do not illuminate their ground targets themselves but record the radiation reflected from other sources (principally the sun) or the target's own self-emitted radiation. In the ultraviolet−visible−near−infrared atmospheric window (0·3 to 1·1 μm) multi-spectral scanners predominantly record reflected solar radiation. In the thermal infrared windows at wavelengths longer than 4·5 μm, and in the even longer wavelengths of the microwave region, the bulk of the radiation is self-emitted and imaging is therefore independent of daylight. In the middle infrared window (3·5 to 5·5 μm) self-emitted and reflected solar radiation may be in approximately equal proportions in daylight. In this waveband day- and night-time imagery of the same target may be quite different in appearance.

The radiation collected along a scan line is focused on to a detector by the optical system. To record the desired spectral information the detector may be specific to a particular narrow waveband; alternatively, wide-band detectors can be used if the radiation is first passed through selective interference filters or separated into its various components by differential refraction using prisms. The detector translates the received (incident) radiation into an electronic signal from which a permanent record of the energy levels comprising the scanned image can be extracted. In the simplest equipment the signals corresponding to each line on the ground are displayed as lines of varying brightness on a cathode ray tube (c.r.t.) These may be recorded from the tube on to a film strip that is moved forward for each successive line to give a permanent ('hard copy') record of the image of the complete swath. (Fig. 5.2). These records on film negative can be reproduced photographically in the normal way. The convention adopted is that high levels of radiation ('hot' targets) appear bright on the positive and low (cold) levels appear dark.

An alternative to direct image recording in the air is to store the image data on magnetic tape for later processing on the ground. This has the advantage of allowing the more complex processing of the signal that is necessary to produce high-quality rectified imagery. Magnetic tape has the advantage that it can store the full dynamic range, that is the complete record of the variation of the strength of the received signals. Photographic films, on the other hand, may be under- or over-exposed at high or low signal levels. Films also have a limited capability for separating large numbers of discrete density levels corresponding to varying signal strengths. This limited 'grey-scale resolution' can considerably reduce the resolution of imagery.

Scanners that record their information on tape generally have some form of additional rapid processing system to provide a 'quick look' at the acquired records. This provides low-quality pictures to show whether proper ground coverage has been achieved and whether the

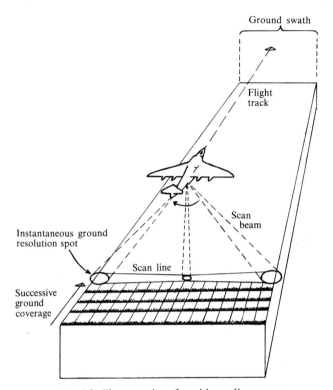

FIG. 5.2. The operation of an airborne linescanner.

signal balance is correct. C.R.T. displays showing the ground coverage one line at a time, or full-frame television pictures successively incremented along the swath, can also be used for in-flight monitoring.

Linescan imagery differs considerably from conventional photography in its geometry. Because of the way in which the picture data are collected the normal principles of perspective geometry do not apply. For this reason linescan strip imagery cannot be examined stereoscopically in the normal way. As the aircraft is moving forward during each lateral sweep of the optical collector system, the image of the scanned swath is not rectilinear in the first instance. Each scanned line is in fact at a slant, the angle of which increases with the distance from the aircraft, giving a sinusoidal distortion to the image. The changing angle of view across each sweep also causes the image to be compressed at the edge of each image strip. The geometry of the scanning spot also varies with its position in relation to the aircraft. The characteristic distortions are illustrated in Fig. 5.3.

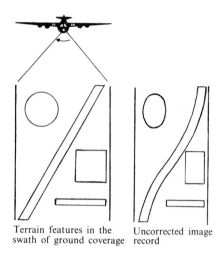

Terrain features in the   Uncorrected image
swath of ground coverage   record

FIG. 5.3. Linescan image distortions.

These distortions are unavoidable with simple direct print-down recording but they can be corrected to some extent by electronic processing, particularly if the data are recorded on magnetic tape. Additional distortions that may arise because of aircraft instability, pitch, and roll may be reduced by mounting the scanner on a stabilized platform or corrected by means of an internal gyroscope system which compensates for roll by altering the position of the recorded scan lines in relation to each other electronically.

The spatial resolution of a linescanner is determined in the first instance by the optical collector system. The size of the collecting surface, or aperture, of the system also, however, affects its spectral sensitivity. A large optical surface will give a high spectral resolution but a low spatial resolution. It follows, then, that the more sensitive the detector element and the quicker its response time the greater the capacity to achieve high resolution.

The size of the ground spot that the optical system can view at any instant is referred to as the *instantaneous field of view* and its size is the *instantaneous ground resolution* of the system. Normally the instantaneous ground resolution of an airborne scanner is not less than 1 milliradian; i.e., when the aircraft is flying at 1000 m the ground spot that is viewed directly under the aircraft is 1 m across. On either side of the flight track the diameter of the instantaneous field of view increases, and its shape becomes distorted as the distance from the optical collector system increases (Fig. 5.2).

As has been indicated, the spatial resolution of a scanner system is adversely affected by slow response in the detector element. This is particularly so with thermal detector elements, which may not be able to register fully the temperature of a particular target spot before being affected by the next target along the scan line. This has the effect of reducing the dynamic range of the recorded data by flattening out the spectral response curve along each scan line so that very hot or very cold objects are not fully recorded. Further, the high response from a hot object may 'carry over' and swamp the low response from an adjacent cold object. This can give a smeared appearance to the imagery.

The signals from the ends of the scan lines at the edges of the image strip have a longer path through the atmosphere than those from directly below the aircraft. Atmospheric absorption and scattering therefore affect the outer edges of the image strip more than the centre. To minimize these effects it is customary to limit the angle of scanning to 90° (45° on either side of the vertical). Some users recommend that the angle should not exceed 70° to give consistent results across the strip.

The performance specifications for a number of scanners are given in Table 5.1. Specific types are described below.

Infrared scanners

Thermal infrared scanners are of the simplest type and were the first to be developed. These instruments use thermosensitive transducers to detect infrared radiation. Typically they use indium antimonide for the $3-5$ μm atmospheric window (Fig. 5.4) and mercury-doped germaniun, lead$-$tin telluride or mercury$-$cadmium telluride for the $8-14$ μm window. Detector elements have to be supercooled to temperatures as low as 77 K by means of liquid air, nitrogen, or helium, according to the equipment. High-performance scanners have black-body reference sources built into them to enable the image data to be correlated with a standard temperature reference. These reference sources work by allowing the detector element to 'see' into a non-reflecting cavity between each scan line while the optical collector system is 'looking' away from the ground.

With some infrared scanning systems the detector elements can be interchanged for different missions to give a choice of waveband. Others are dual-channel instruments with separate collector systems for each detector, one at each end of the instrument.

Table 5.1. Some thermal and multispectral scanner specifications

Equipment	Channels	Spectral range (μm)	Angular resolution (milliradians)	Across-track field of view (scanning angle) (degrees)	Collecting aperture (cm)	Scan rate (oscillations per second)
EMI Airscan	1	3·5–5·5 or 8–14	1·5 or 3·3	140	10	50
Hawker-Siddeley 212	1	8–14	1·5	120	9	500
Texas Instruments RS–310	1	Selectable 0·3–14·0	1·5	90	16	200
Texas Instruments RS–14	2	Selectable 0·3–14·0	1·0 or 3·0	80	16	200
Daedalus	12	0·3–14·0	1·7 or 2·5	77	13	80
Environmental Research Institute of Michigan	12	0·41–9·82	3·0	80	–	–
Bendix M²S	11	0·38–11·0	2·5	100	10	10–100
Bendix MSDS	24	0·34–13·00	2·0	80	23	10–100

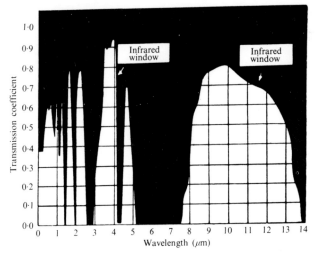

FIG. 5.4. Thermal infrared windows.

Multispectral scanners

Multispectral scanners can operate from the ultraviolet to infrared but most of their channels are usually in the visible band. The radiation collected by these instruments is split optically and each part is simultaneously recorded by a separate detector element. Scanners have been built with up to 24 separate channels but 4 to 14 channels is a more usual number. Table 5.2 lists the separate wavebands recorded by a 24-channel instrument.

Multispectral scanners generally use photo-emissive silicon detector arrays for the visible and near-infrared parts of the spectrum. These are photomultipliers and have faster reaction times than thermal infrared detectors.

Microwave scanners

Microwave scanners are essentially radio receivers for super- to extra-high-frequency wavelengths. They are normally referred to as *passive microwave radiometers*. Their aerials are adapted to a line-scanning mode. They operate in the 1- to 30-mm waveband corresponding to a frequency range of 300–10 GHz.* There are some atmospheric absorption bands in this range but the atmosphere becomes increasingly transparent towards the longer wavelengths (Fig. 5.5).

The instruments record the natural radio emissions from the earth's

* GHz = gigahertz (1 hertz = 1 cycle per second).

Table 5.2. Bendix MSDS multi-spectral scanner: spectral bands

Channel	Bandwidth (µm)	Detector	
1	0·34 – 0·4	Photomultiplier	Ultraviolet
2	0·4 – 0·4		
3	0·46 – 0·5		Blue
4	0·53 – 0·57	Silicon	Green
5	0·57 – 0·63		
6	0·64 – 0·68		Red
7	0·71 – 0·75		
8	0·76 – 0·80		
9	0·82 – 0·87		Near-infrared
10	0·97 – 1·05		
11	1·18 – 1·30	Germanium	
12	1·52 – 1·73		
13	2·1 – 2·36	Indium antimonide	
14	3·54 – 4·0		
15	4·5 – 4·75		
16	6·0 – 7·0	Mercury-doped germanium	Middle infrared
17	8·3 – 8·8		
18	8·8 – 9·3		
19	9·3 – 9·8		
20	10·1 – 11·0		
21	11·0 – 12·0		
22	12·0 – 13·0		
23	1·16 – 1·16	Germanium	Near-infrared
24	1·05 – 1·09	Silicon	

surface, of which the amplitudes are several orders of magnitude lower than for emissions in the middle (thermal) infrared. To be able to detect such low levels of radiation and separate them from the radio 'background' the aerials must have wide angles of view – of the order of degrees rather than the milliradians customary for thermal scanners. As a result, the spatial resolution of these instruments is inherently very low.

The main components of a microwave scanner are the aerial or antenna and a wide-band receiver tuned to a selected 'centre frequency'. The aerial may be a conical horn lens or a parabolic reflector and may itself rotate or have a rotating scanning plate mounted in front of it. An alternative method employs electronic scanning; a fixed arrangement of antenna elements is phased so that each element records a segment

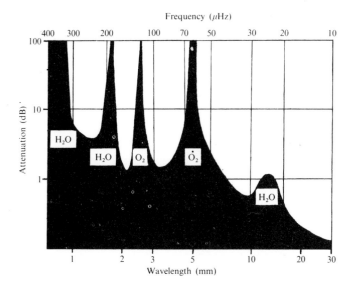

FIG. 5.5. Part of the microwave spectrum, showing absorption by atmospheric constituents and 'windows'.

of ground information in turn in a pattern simulating a line scanner. This is known as a 'phased array'.

For image recording, because the resolution is so poor, high-quality image output is not essential. A simple graphic computer-type printout may be produced or the image can be recorded on tape and displayed on a video screen. In either case it is possible to code different levels of signal strength by colour or symbol to produce images that are in fact contour maps of microwave temperature similar to the 'heat maps' that can be similarly produced from thermal infrared mappers.

Table 5.3 gives details of some types of equipment. It will be noticed that the size of the antenna and the spread of the scanning beam increases with increasing wavelength (decreasing frequency). This means that heavier and bulkier systems are necessary for longer wavelengths.

Scanner operations

Infrared and multispectral scanners are only slightly more difficult to operate than aerial survey cameras. The basic scanning units are quite small; they can be mounted in the standard camera hole in a survey aircraft, or they can be carried outboard on a light aircraft. The tape recorders and monitoring equipment carried inside the aircraft are not particularly bulky and the cooling necessary for the detector elements of thermal scanners does not create any serious difficulties. The large

Table 5.3. Some airborne microwave radiometer systems

Equipment	Centre frequency (GHz)	Beam width (Angular resolution) (degrees)	Temperature resolution (K)
McDonnel–Douglas	10·7	20·0	1
Spectran RA 111	30·0	2·4	0·21
Spectran RK 112	13·7	3·2	0·15
Spectran RX 112	10·2	4·3	0·11
Spectran RS 112	2·4	10·0	0·15
DFLVR	11·0	1·88	0·1
DFLVR	32·0	5·0	0·5
DFLVR	90·0	1·8	3·0
DFLVR	32·0	1·0	20·0

antennae of some microwave scanners may pose some problems but the other elements of these instruments are not large. Flying operations with scanners are similar to those for aerial photography except that night flying, which is often required for infrared and microwave surveys to eliminate the confusing effects of direct solar reflectance, may call for more sophisticated navigation aids such as doppler equipment.

Applications

Photographic cameras produce images of better planimetric accuracy and spatial resolution than scanners. The chief advantage of scanners over cameras is that they can collect information in spectral channels that cannot be recorded directly on film and some of them can record information simultaneously in a greater number of discrete channels. Further, the spectral bands can be more precisely defined and the spectral data more readily and reliably quantified.

There has been considerable use of data from multispectral scanners in the United States, notably at Purdue University and the University of Michigan, for mapping land resources automatically by computer processing. These techniques, in which different mapping units are recognized by their spectral signatures, are discussed in Chapter 8. Most of this work has concentrated so far on the mapping of crops and natural vegetation because large uniform areas of crop or natural vegetation offer relatively easy targets for this approach. There has, however, also been some work on geology, land-forms, and soil.

In one instance the Laboratory for Applications of Remote Sensing (LARS) at Purdue University analysed multi-spectral scanner data for an area and produced maps of 'spectral classes' of closely related spectral signatures equivalent to three soil classes as compared to five classes mapped in the field. In fact, the spectral classes corresponded more closely to land management groupings than to soil units *senso strictu*. The overall conclusion from this work was that in this particular test area a one-to-one correspondence between the mapping units used for conventional soil survey and spectral classes was unlikely owing to the gradation between soil types and the lack of definite boundaries. (Cipra et al 1972).

This example demonstrates one of the main problems of this kind of approach to soil mapping. Though the soil colour is one form of signature, it can be accurately sensed only if the soil is clear of vegetation. Furthermore, the signature of a soil body refers in general only to its surface, which may or may not be directly related to any other more important feature of the profile. It is the boundaries between surface expressions that are most useful in mapping, but these surface expressions are not necessarily either uniformly important or particular to the underlying soil.

Single- or dual-channel thermal infrared scanners have fewer applications to conventional soil survey than multi-spectral scanners. With most of the instruments available, low resolution and image distortions, which make it difficult to match adjacent strips of cover, tend to outweigh the advantages of having thermal infrared information. The infrared, however, may give information on soil texture, moisture distribution, and possibly on the mineral composition of surface materials that could be of interest for special soil surveys for engineering purposes or possibly for land management.

Geologists have shown some interest in the use of the *Restrahlen effect* to identify rock-types. In the $8-14$ μm middle-infrared window the level of minimum emission from basic rocks (i.e. those with low silica contents) occurs at much longer wavelengths than for 'acid' (silica-rich) rocks, provided that both the materials are at the same temperature. By comparing image data in two channels centred on 9 μm and 11 μm wavelength, it is possible, in the right circumstances, to discriminate between the two rock-types automatically. This technique could be of value for some types of soil survey, particularly in arid areas (Lyon 1965; Vincent and Thomas 1971).

Microwave instruments, unlike scanners operating in the thermal and visible bands, can record information from below the immediate ground surface. This opens up the possibility of mapping subsoils by remote sensing, but in fact the radiation is so heavily attenuated by moisture in the soil that contributions from any appreciable depth are detectable only through the driest of materials. In theory penetration of several tens of metres is possible through dry sand; and the longer the wavelength used, the greater the depth from which radiation can be detected.

For more general applications, the fact that microwave sensors are not affected by cloud cover could make them useful for regular monitoring of soil moisture conditions for agricultural management. The low resolution of the instruments means that regional rather than detailed coverage would be most appropriate.

There has been considerable research effort into the use of scanning imagers of all kinds, but few practical applications have emerged so far, except for the widespread use of small-scale imagery from the multi-spectral scanners of the *Landsat* (Earth Resources Technology Satellite) series (discussed in Chapter 7). Certainly the use of airborne scanners for soil mapping work has been entirely experimental. Lack of progress so far may be partly due to the limited opportunities for those working in the field to use material of this kind. However, its inherent drawbacks in image resolution and geometry compare unfavourably with aerial photography. How far improved instruments and data processing, and increased availability of scanner imagery, will benefit soil survey remains to be seen.

6. Side-looking radar

Radar systems with fixed or rotating aerials were first developed for day or night military surveillance and for navigation in all weathers. (The word 'radar' is derived from 'radio detection and ranging'.) Side-looking radar (SLR) from aircraft was developed later to produce permanent image records of the ground analogous to aerial photography. Like all radar systems, and unlike *passive* microwave radiometers, it operates in an active mode, that is it emits its own radiation and records the signals reflected back from the ground target.

Radar uses radio wavelengths, as the name implies, from 3 mm to 3 m (100 GHz to 100 MHz). Wavebands within this range are given letter designations for reference, but some confusion arises because the United States, Great Britain, and NATO use different codes (Fig. 6.1). Side-looking radar mapping systems have been built to operate in the Ka, X, S, and L bands of the American designations.

Side-looking radar works by directing a narrow beam of pulsed microwave energy from a fixed aerial on the aircraft at right-angles to the line of flight. This illuminates successive strips of ground to make up a continuous swath of coverage similar to that produced by a scanner. The radar echoes from each strip or line are received by the same aerial and the variations in the strength of the signals reflected from different segments of the line are used to construct an image of the terrain (Fig. 6.2).

The picture is recorded by a technique similar to that used to print-down linescanner imagery: the signals from each line modulate an electron beam which traces a line of fluctuating brightness across the face of a cathode ray tube. In the simplest equipment the successive lines are recorded directly on moving film to build up the continuous strip image. The distance from the aircraft, or the 'range', of each segment in a strip is determined by the time delay between the emission of each pulse and its return signal, and this information is used to reconstruct the correct geometry of the picture. The width of the swath of coverage is determined by the angle of depression (from the horizontal) of the radar beam. This is typically 15°. Beneath the aircraft the angle from the vertical of the inner edge of the beam is not normally smaller than 5°. This is because at high angles of incidence so much of the signal returns from the target that it saturates the recording, with consequent loss of contrast and detail in the image. Typical SLR systems obtain coverage with swath widths of 10 to 40 km from flying heights of between 3000 and 7000 m. Some military equipment obtains simultaneous coverage on each side of the aircraft, omitting a strip immediately under the aircraft for the reasons already explained.

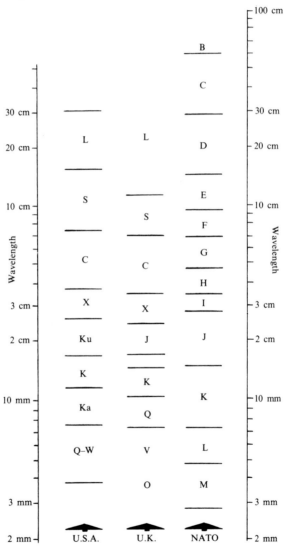

FIG. 6.1. Radar waveband nomenclature.

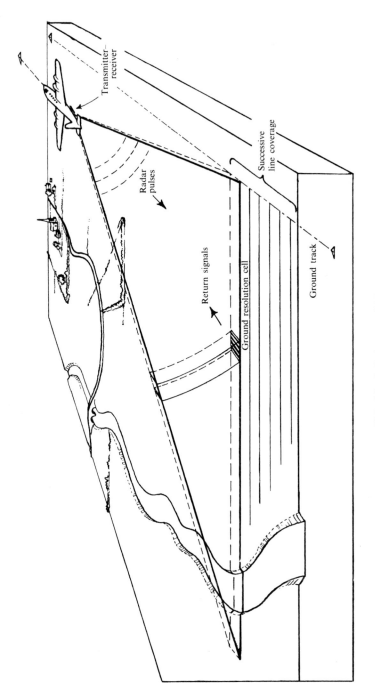

FIG. 6.2. SLR data acquisition.

Commercial SLR systems, however, generally use a single 'look direction' for a mapping survey, because the illumination should always be in the same direction for adjacent strips to maintain consistency in image appearance.

To obtain imagery with high planimetric accuracy a number of corrections have to be made to the image data. In an uncorrected image the scale decreases with increasing distance from the aircraft across the swath, as with oblique photography. High-quality mapping radars correct this 'slant range' imagery by manipulating the displayed signals to achieve an approximately even scale across the width of the full swath (Fig. 6.3). Doppler or inertial navigation systems provide accurate position-fixing to compensate for any scale errors in the along-track (y) direction that may arise from variations in aircraft speed. The angle at which the individual lines of the image are recorded can be altered to compensate for aircraft yaw, and some systems use stablized aerials to compensate for roll.

The target resolution of an SLR system is initially determined by three factors: the wavelength employed, the length of the pulse of energy, and the width of the beam. Since in theory imaging radars cannot detect single objects of dimensions smaller than that of the radar wavelength, short-wave radars should have potentially higher resolving power than longer-wave systems (Fig. 6.1). In practice, other constraints override this.

The length of the microwave pulses, which may vary from several microseconds to 10 nanoseconds, is determined by an electronic clock known as the 'time base'. The cross-track or range resolution of the system (that is, the resolution in the x direction) is determined by this; the shorter the pulse the higher the resolution. By using very short pulses some military radars appear to be capable of resolutions down to 25 cm, which is about 30 times better than the radar mapping systems available for commercial or civil use.

Along-track resolution in the azimuth or y-direction, is determined by the width of the transmitted beam (or the part of it from which the picture is generated). In the simpler versions of SLR, known as 'real-aperture systems', the beam width is determined by the length of the antenna. The larger the antenna the narrower the beam:

$$(\text{Beam width}) = \frac{(\text{Wavelength})}{(\text{Antenna length})}.$$

Ideally the pattern of radiation from the antenna should be in the form of a fan. In practice it is lobed, and if the length of the antenna is increased, the central lobe is elongated until it becomes effectively a narrow fan-beam (Fig. 6.4). 'Real-aperture' radars use antennae from

1·5 to 5 metres in length to give reasonable resolution in the y-direction. As the beam widens away from the aircraft, resolution in this direction decreases across the swath of coverage. This is not, however, the case with the cross-track resolution, which is, as we have seen, determined by the pulse length.

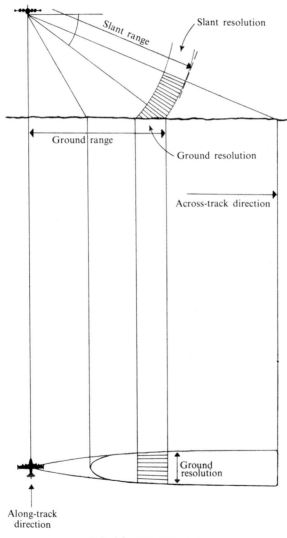

FIG. 6.3. SLR geometry.

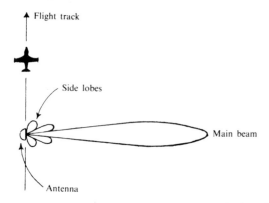

FIG. 6.4. Radar antenna pattern. This diagram is a plan view of a three-dimensional configuration. The side lobes show the same pattern in the vertical plane.

The strongest return signals come from ground objects with faces perpendicular to the radar beam. Even if they are much smaller than the nominal resolution of the radar, highly reflecting objects of this sort will still be detected because of their high signal returns. For practical purposes, however, the resolution of a radar system is defined as the smallest separation distance it can detect between objects of equal reflectance.

As SLR is an active system the radio waves can penetrate the ground or substrate that they 'illuminate' to varying degrees according to the power used, the wavelength, and the nature of the substrate. In general the longer the wavelength the greater its penetrating power, but the degree of penetration depends greatly on the medium. The atmosphere is almost completely transparent to most of the microwave-radar spectrum and most wavelengths easily penetrate cloud and mist, which makes mapping possible in all weathers. Falling rain and snow scatter and absorb some radiation, particularly in the shorter wavelengths. The microwaves achieve some penetration into soil but the extent seems to depend more on the soil moisture content, which heavily attenuates the degree of penetration, than on the mineral composition. Open water surfaces reflect almost all the microwave radiation they receive and there is practically no penetration.

Most side-looking radars are monochromatic, that is, they operate with pulses of a single wavelength. Because of the highly specific responses of the various facets of a target area the ground appears to glitter when illuminated by microwaves. This produces characteristic speckled patterns on some imagery from high-resolution side-looking radars. It has been suggested that panchromatic or 'frequency-agile' radars using a narrow waveband rather than a single wavelength could smooth out the pattern of response to give images of better quality.

Multi-frequency and multi-polarized systems have been exploited to give additional target information. Multi-frequency systems use two or more frequencies and produce separate image records for each wavelength. The polarization of the radar emission is determined by the design of the antenna. Many ground targets give different levels of response to different polarizations. The principal side-look radar systems transmit horizontally polarized radiation and record the part of it that returns with its polarization unmodified by interaction with ground targets. This is known as HH recording. Some radars, however, have two channels which combine a horizontally polarized with a vertically polarized return record; this is termed HV recording (Pl. 6).

Some SLR systems use what is known as the *synthetic aperture* technique in order to obtain high resolution without having to use very large antennae. They employ a small antenna with a widely spreading beam, but the returning signals are filtered so that the image is produced from data returned from only a narrow part of the beam. The technique takes advantage of the fact that the forward motion of the aircraft induces Doppler shift in the received signals which increases the frequency of those returning from ahead of the aircraft and decreases it in those from behind. The signals are filtered electronically so that only those from a narrow segment in the centre of the fan-beam are used to generate the image (Fig. 6.5). Image resolution can be further improved by electronic 'focusing', again using the Doppler shift effect, in which case the resolution can be made independent of the distance from the target (see Table 6.1). Synthetic aperture radars can potentially give a resolution ten times better than that of real-aperture systems, but they require much more complex data-processing units.

SLR is available for civil uses with both real and synthetic aperture; Table 6.2 lists equipment used up to 1974. The Remote Sensing

Table 6.1. Comparative resolutions of radar systems.
Theoretical resolutions at 10km range

Wavelength (mm)	Waveband	Real aperture 5-m antenna (m)	Unfocused synthetic aperture (m)	Focused synthetic aperture (m)
8	K	16	8	4·5
30	X	60	16	4·5
300	L	600	154	4·5

Table 6.2. Radar systems

	Wavelength (mm)	Waveband*	Nominal resolution		Polarization	Maximum swath width (km)	Aircraft
			Azimuth at 15 km (y direction) (m)	Range (x direction) (m)			
Westinghouse real aperture. APQ-97	8·6	Ka	25	8	HH,HV	21	DC-6B
Goodyear synthetic aperture. APQ-102	31	X	15	15	HH	37	Caravelle
Motorola real aperture. APS-94(D)	30	X	116	30	HH	100	Mohawk and Gulf-Stream
Environmental Research Institute of Michigan Dual-channel synthetic aperture system	31 and 300	X L	10 10	10 10	HH,HV HH,HV	18 18	C46

* United States designations

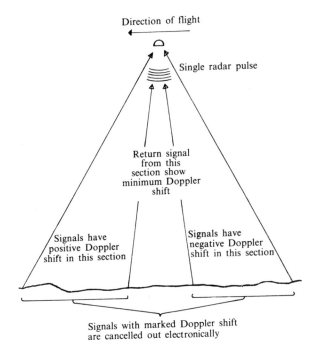

FIG. 6.5. The generation of a synthetic aperture.

Laboratory of the University of Kansas is experimenting with a cheap
and simple real-aperture X-band system designed to be carried in light
aircraft. The image is recorded by photographing a direct video display
and the potential resolution is 20 m.

SLR imagery

Although SLR imagery is produced in photographic format and
could be mistaken for aerial photography it has some special characteristics. The scale at which the image is initially recorded depends on the
ratio of the width of the swath to the width of the print-down. Unlike
aerial photography the resolution is independent of the acquisition
scale. Initial image scales vary with the equipment; for instance, the
Westinghouse real-aperture APQ–97 system normally uses a print-down
scale of 1 : 225 000, whereas that of the Goodyear APQ–102 synthetic
aperture system is 1 : 400 000. Both these systems have the same
nominal resolution.

As the radar image of a swath is built up line by line, it is not possible
to obtain stereo-cover from a single pass as with aerial photography.

The strong hill-shading effects, which result from the oblique view of the terrain, compensate for this to some extent. Stereo-cover can be obtained by flying adjacent swaths of imagery with a lateral overlap. The two strips can then be viewed under a stereoscope in the normal way. Unfortunately the stereo-images usually have variable distortions, for the matching of detail between strips is not precise, and the quality of the print-downs is far below that of aerial photographs.

Because the view in SLR imagery is oblique there are considerable areas of shadow, particularly in rough terrain. Shadows on photographs are areas from which insufficient reflected light has been obtained to expose the film; with radar they are areas of no information at all, the microwave signals having been blocked by an intervening object so that the area is recorded as a blank on the print-down (Fig. 6.6). These shadows lengthen with distance from the source of illumination, and it is possible to make approximate measurements of terrain height from the length of the shadow, the angle of incidence of the radar beam (which can be calculated from the height of the aircraft), and the position of the image in the swath.

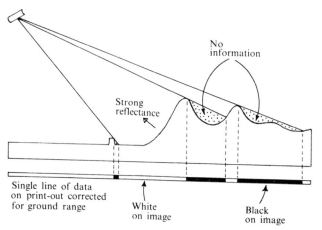

FIG. 6.6. The radar shadow effect.

Because the timing of the return signals from a target is used to determine its distance, steep objects present particular problems. The signal reflected from the top of a high building, for instance, will return before the signal from its base and the image will be inverted on the record. In the same way, the images of steep slopes facing the source of radar illumination will appear to lean over. These distortions are less severe on imagery that has not been corrected for ground-range, and some interpreters prefer to avoid ground-range correction for this reason.

As the shadowing effects in radar imagery give a strong impression of relief, imagery should always be viewed with the apparent source of illumination at the top of the picture; the relief may otherwise appear to be inverted. Because of this, it is customary to fly coverage of large areas on east−west flight lines with the illumination from the south. The shadowing then corresponds to the hill-shading convention of cartography. It may be necessary to use alternative look-directions in mountainous areas to minimize confusing terrain effects.

The look-direction of the system can greatly affect the interpretation of the imagery, since terrain features may be enhanced or suppressed by shadows. Linear features at right-angles to the look-direction may, on the other hand, be particularly difficult to interpret; in steep terrain they may be obscured by shadows.

Although photogrammetric accuracies cannot be obtained even with high-quality SLR systems, planimetric corrections can be made for scale changes across and along the track and for aircraft yaw, and mosaic sheets can be produced that are as accurate as semi-controlled photo-mosaics. A typical specification for radar mosaics would require that the cumulative scale discrepancy in any direction should not exceed 1 km, the angular distortion should not exceed 10 milliradians, and the corner positions of each mosaic should be accurate to within 1 km with a probability of 95 per cent and to within 0·5 km with a probability of 50 per cent (Innes 1972). This is for 1:250 000-scale material.

As mentioned above, the monochromatic illumination may impart a characteristic speckled pattern to the imagery. This is most marked in focused synthetic aperture systems of high resolution, and may limit the size to which individual negatives may be usefully enlarged. The relatively low resolution of the image may also influence this but in practice it has been found quite feasible to carry out interpretation and mapping on materials that have been enlarged as much as five-fold over the acquisition scale. Standard negatives from the Goodyear synthetic aperture radar, originally at 1:400 000 scale, may be enlarged to 1:250 000 scale mosaics, and Westinghouse real-aperture imagery, initially at 1:225 000 scale, can be readily enlarged to a scale of 1:50 000.

Commonly the imagery from real aperture radar shows alternate darker and lighter bands running parallel to the flight direction and most strongly expressed in the near-range area (Pl. 6). These are due to the antenna lobing (see Fig. 6.4), which produces different signal strengths in different parts of the swath.

Very flat surfaces like water or roads reflect practically all the radar energy away from the aircraft and such targets register as black: a nil return. In contrast, any relief or surface roughness may be revealed very strongly. In very steep terrain, however, the surface roughness of individual facets on slopes facing the direction of illumination may be obscured by high overall signal returns, and on the reverse slopes by shadow. In flatter terrain the glancing angle of the beam allows a multiplicity of micro-facets to be resolved. In general, surface roughness, resulting from micro-relief and vegetation, and the effects of variations in moisture content in the substrate and in the vegetation together give a highly sensitive picture of the ground surface which can be of great value in soil mapping.

The use of SLR

Side-looking radar coverage is a useful substitute for aerial photography in regions of prevailing bad weather where medium-scale coverage of large areas is required. The most important property of the system here is its ability to obtain imagery through cloud. On the other hand, its oblique view and its sensing of the ground at particular microwave frequencies may be of specific value in natural resources surveys. Much work remains to be done in order to assess the full potential of SLR in this field.

SLR has been used for mapping natural resources since the late 1960s. The main applications have been in mapping the geology of concession areas for oil and mineral exploration but SLR is being increasingly used for mapping other resources and for basic cartography. For example, Project Radam in Brazil produced complete coverage of the greater part of the Amazon basin for the first time. This has been published as mosaic sheets at scales of 1:250 000, 1:100 000, and 1:12 000, and has provided the first systematic and accurate base maps of the entire area. The mosaic sheets have been used as base maps and interpretation material for surveys of all kinds of natural resources including semi-detailed soil surveys. (de Godoy and Van Roessel 1972). Another example is the SLR survey of Nicaragua carried out in 1971. This covered the whole country and the imagery was used to produce reconnaissance maps of the geology, geomorphology, vegetation and land-use.

One fact is clear: SLR imagery is particularly valuable for mapping landforms. Its main immediate contribution to soil mapping lies in the delineation of geomorphologically defined soil associations or land systems.

Practical work has shown that soil units may also be delineated by reference to the distribution of vegetation types which show up clearly on radar imagery, particularly in terrain with low relief (Pl. 7b). Turgid swamp vegetation, for instance, gives strong signal returns and may be readily separated from drier vegetation types by its bright signature on the image. Mature woodland, on the other hand, having facets consisting of individual tree canopies, has a strongly and uniformly mottled appearance on SLR imagery compared to the darker and more finely varied tones characteristic of grasslands.

It is difficult to assess how far radio waves have any specific value for soil survey. It is not really known how far microwave radiation will penetrate soils under varying field conditions, nor how such data could be used to help in mapping. Certain soil types do, however, have a characteristic appearance on radar imagery. Dry sands in particular give high signal returns because of high interior reflectance in the individual silica grains, and they thus appear bright on the image. The possibilities of more specific spectral responses than these are, however, limited by the monochromatic illumination of radars. The use of different polarities for transmitted and received radiation may give additional information on some soil types but there is so far little evidence on this.

Perhaps soil depths over large areas could be mapped by means of radar but there has been no practical demonstration of this so far. What is more immediately relevant to soil survey is that the longer-wave radar systems being experimented with may be able to penetrate through vegetation cover to reveal otherwise obscured details of the land surface.

Operations and costs

Because of its relatively low resolution and high cost, SLR mapping is most suitable for the coverage of large areas. The rate of survey is very rapid. For example, in 1971 the whole of Nicaragua, totalling 148 000 km², was covered in 9 days by a single aircraft equipped with a Westinghouse real aperture system flying at 6000 m. SLR imagery of 4·5 million km² of the Amazon Basin was obtained for the Radam Project by a jet aircraft equipped with a Goodyear synthetic aperture system flying at 13 000 m in two flying seasons (1971 and 1972).

Such large-scale operations require aircraft that can remain in the air for long periods to produce as much coverage as possible on each mission. In areas subject to bad weather much of the flying is 'blind'. The aircraft must be fitted with advanced inertial or Doppler automatic nagivation equipment. The aircraft used for the Radam Project, for instance, employed an inertial navigation system cross-referenced by a SHORAN radio position-fixing system, plus a radio altimeter. Such operations are highly expensive. If, however, the projects are large enough the costs of a large aircraft and its crew, and of sophisticated equipment and data-processing, are offset by the rapid rate of acquisition and a guarantee of coverage even in very bad weather. It has been claimed that SLR coverage for large areas is no more expensive per unit area than conventional aerial photography. However, because of the high mobilization charges for aircraft, crew, and equipment it is probably not economical to fly areas of less than 50 000 km^2, though a number of smaller areas could be combined in one operational tour if they amounted in total to a comparable area. Not many soil mapping projects will be large enough to justify specifically commissioned SLR coverage. On the other hand, coverage originally taken for other or more general purposes is available for large parts of the world, mostly in the wet tropics.

Future prospects

The relatively low resolution of current SLR systems and the limited planimetric accuracy of their imagery restrict their use to reconnaissance or at best to semi-detailed soil mapping. There are prospects of much better ground resolution, of higher planimetric accuracies, and of refinements such as multi-spectral and multi-polarized coverage. Usable ground penetration systems may be developed and all of these may prove to be applicable to soil work. We cannot, however, expect to see routine widespread use of SLR in the near future.

Further in the future an interesting prospect is for a SLR system to be mounted in a space satellite. Resolutions of 20 to 50 m are possible from satellite altitudes, using synthetic aperture systems, and with such vehicles complete world coverage could be obtained in a relatively short space of time. They would also give opportunities for regularly repeated coverage regardless of weather conditions. A number of prospects for this are being examined for the 1980s, including automatic satellites and manned space stations. However such coverage is obtained, it will be used for soil and soil-related mapping and monitoring projects of one sort or another once it is made available.

7. Imagery from space platforms

The use of cameras and other types of imaging device from orbital altitudes is an obvious application of rocket and space research programmes. In early work before 1950 cameras were carried in rockets but this was more to record the trajectories by reference to ground detail than to obtain pictures for the study of earth resources. Later the manned *Gemini* and *Apollo* satellites produced large numbers of photographs of the earth. Unmanned satellites taking pictures with automatic sensing devices were first developed for weather forecasting and the value of this technique for military surveillance was rapidly appreciated. Satellites designed specifically for the mapping of earth resources have been a relatively late development. An interest in the subject is now, however, well established and photography specifically taken for investigations into earth resources has been obtained from the manned American *Skylab* missions (1973) and the Russian *Salyut* missions (1974−5).

Because of the distances involved, imagery from orbital altitudes is inevitably of small scale and low resolution compared to that of aerial photography. An individual frame covers a very large area, and this 'synoptic' characteristic is probably the most important feature of space imagery. Because of this, improvements in sensor resolution which could mean a reduction in the overall field of view are considered by some to be inappropriate. The higher the resolution of the sensor, the greater the quantity of data in the picture to be transmitted by radio link from an automatic imaging satellite. So with satellites of this type the concept of 'maximum desirable resolution' arises. The problem is not so severe with manned missions when film records can be physically recovered.

Another important factor is that satellite-borne sensors with wide fields of view can cover very large areas of ground very quickly and can repeat their coverage at regular short intervals. Satellites are thus ideally placed for monitoring changes in land-use, vegetation, or soil moisture conditions.

Satellite orbits vary in angle of launch, period (the time taken to complete a circuit of the earth), speed, and altitude. The orbits of most manned missions have so far been inclined at low angles to the equator. For instance, the *Skylab* orbit was inclined at 50° to the equator and so it did not overpass areas north or south of 50° latitudes. Weather satellites which are designed for full Earth coverage are in polar orbits at small angles to the lines of longitude. By precise timing of its orbit a satellite can be made sun-synchronous so that its track always lies in the same position in relation to the 'daylight terminator' or earth's shadow. This means that the satellite will cross the equator (or any other latitude) at the same time each day, though not necessarily at the same longitude. This is of obvious value when repeat imagery is required from sensors in the visible band.

Gemini and Apollo

The *Gemini* and *Apollo* series of manned satellite launchings that culminated in the Moon landings were also used for a variety of other space research projects, including photography of the earth. A number of missions between 1965 and 1969 carried cameras for this purpose; these are listed in Appendix 3. The instruments were mostly 35-mm and 70-mm hand-held cameras using colour, false-colour, and black and white film with various filters. Most of the photographs were high-angle oblique shots at very small scales. Object resolutions as fine as 30 m were achieved in some instances, despite the small film formats and the short focal length of the cameras.

The coverage from the *Gemini* and *Apollo* missions was somewhat random, and the low inclination of the orbits to the equator limited photography to the zone between latitudes 30° N and 30° S. The picture quality was very variable. Many exposures suffered considerably from the effects of atmospheric dispersion and were rather blue in tone. The photographs nevertheless contained a great deal of information.

The value of small-scale synoptic photography was immediately apparent to people making reconnaissance studies of areas that had previously been poorly mapped; for some areas it was the first imagery ever available. The geological and topographic maps of the southern Sahara produced by Dr. Angelo Pesce were one outcome of this new type of coverage. These included the first accurate topographic outline maps of south-eastern Libya and adjoining desert areas in Egypt and the Sudan. An area of dunes some 200 by 40 km in extent lying east of the Tibesti mountains which had not previously been mapped was, for example, interpreted from the *Gemini* photography.

As well as providing broad new topographic and geomorphological information from poorly mapped areas the satellite photographs also provided more subtle information for better-known areas. The *Gemini* photograph of Jebel Marra in the western Sudan (Pl. 4a) illustrates this. The author produced a reconnaissance soil map of this area for FAO in 1964. Two tracts of soil in particular appeared to have potential for agricultural development. These were developed on deep volcanic ash lying due west and due south of the lower slopes of the Jebel Marra volcano with its two prominent crater lakes. The soils had essentially the same profiles in both areas except that the soils in the southern tract contained between 10 and 20 per cent more sand in their surface horizons, making the difference between sandy loams and loams in the top 25 cm. At the time of mapping this difference was attributed to drift material from the siliceous Basement Complex rocks exposed only on the southern slopes of the mountain. The *Gemini* photography clearly indicates another reason: it shows that the bright orange-coloured area to the north-east of the mountain ridge, which ends in the prominent cone of Jebel Marra, is part of the Sudan Goz System — a field of fixed dunes originating from an interpluvial extension of the Sahara. The main sand sheet does not extend west of the mountains, nor immediately south of the mountain. The extension of the orange colour in this area over the southern tract of ash soils indicates a common origin for the dune field and for the topsoils in that area. On the photograph the area of otherwise identical soils west of the mountain where the topsoils are not sandy is bluish in colour.

It is significant that even this essentially random photography of very small scale from orbital altitudes proved to be of value in mapping earth resources. Systematic higher-resolution coverage from special-purpose satellites should therefore be of even greater value.

Weather satellites

The *Nimbus 1* experimental weather-watch satellite was launched from the United States in 1964 as the first in a series now joined by operation Environmental Science Service Administration (ESSA) and National Oceanic and Atmospheric Administration (NOAA) series vehicles in the civil field as well as both United States and Russian military meterorological satellite systems.

Weather satellites are designed to operate for the most part in circular near-polar orbits (at high angles to the equator and thus passing over the poles). They carry a variety of sensors to record atmospheric and oceanographic data for meteorological purposes. Some of these are imaging devices to collect picture data in a range of spectral bands from the ultraviolet to the far infrared (see Table 7.1). The satellites are completely automatic and all data are relayed to ground stations by radio telemetry links. The image data are received from the sensors as fluctuating electronic signals which can be reconstituted in various forms of scanning recorder.

The resolution of all these instruments is lower than desirable for mapping earth resources. This is an inevitable result of the compromise

Table 7.1. Imaging sensors on the Nimbus satellites (1964−70): Area coverage and resolution of the sensors

	Area coverage	best resolution (km)
AVCS: Advanced vidicon camera system (Three 800-line vidicon cameras)		
Nimbus 1	300 × 1200 km (perigee)*	0·33
	650 × 3000 km (apogee)*	0·77
Nimbus 2	750 × 3700 km	0·92
APTS: Automatic picture transmission system (Wide-angle television camera)		
Nimbus 1	850 × 850 km (perigee)*	0·80
	1950 × 1950 km (apogee)*	1·80
Nimbus 2	2200 × 2200 km	2·20
IDCS: Image-dissector camera system		
Nimbus 3 and 4	2700 × 2700 km	3·3
HRIR: High-resolution infrared radiometer (Scanning radiometer)		
Nimbus 1	1800-km swath (perigee)*	3·3
	6500-km swath (apogee)*	7·5
Nimbus 2	7100-km swath (horizon to horizon)	9·3
Nimbus 3	7100-km swath	8·5
Nimbus 4	7100-km swath	7·7
MRIR: Medium resolution infrared radiometer (scanning radiometer)		
Nimbus 2 and 3	7100-km swath	60

*The elliptical orbit of Nimbus 1 was an unplanned aberration. Perigee is the lowest point of an eliptical orbit, apogee the highest.

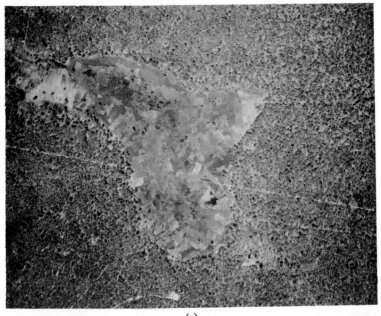

(a)

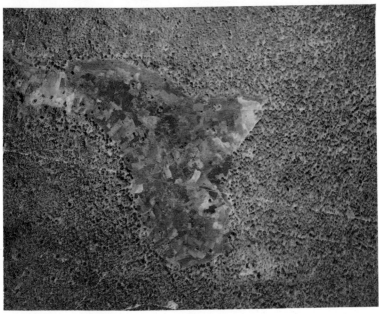

(b)

1. Photography in colour and false-colour infrared
(a) Aerial photograph in true colour of an area of cultivation in *Brachystegia* bush country, Zambia.
(b) The same site on infrared Ektachrome film. The details of cultivation are much clearer and the contrasts between tree types are more marked.

2. Multispectral photography

Photograph from Southern Africa, taken by a four-lens multispectral camera (International Imaging Systems). The channel numbers on the outer corners of the frames are: channel 1, blue; channel 2, green; channel 3, red; channel 4, near-infrared. There are marked differences in the detail and contrast of vegetation, cultivation, and land-forms between the four channels.

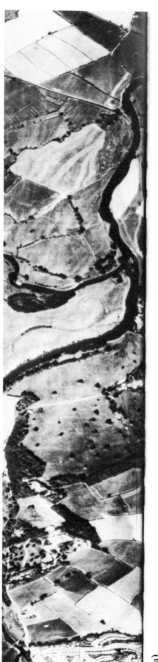

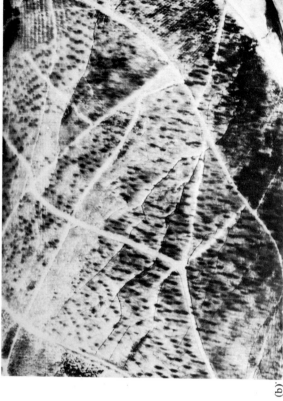

(a)

(b)

3. Thermal infrared imagery

(a) Strip imagery along the River Trent taken in the 3·5-5·5-μm band by EMI Airscan equipment, showing differences in temperature between fields and within fields due to variations in the nature of the surface.

(b) Part of a forest plantation recorded in the 8-14-μm band by a Hawker Siddeley Linescanner. The tracks, which are warm, show white; the ditches carrying water (cold) are black; ditches without water are much fainter.

(a)

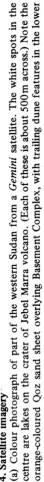

(b)

4. Satellite imagery

(a) Colour photograph of part of the western Sudan from a *Gemini* satellite. The white spots in the centre are lakes on the crater of Jebel Marra volcano. (Each of these is about 500 m across.) Note the orange-coloured Qoz sand sheet overlying Basement Complex, with trailing dune features in the lower right.

(b) False-colour composite frame of part of central Iran, taken by the *Landsat* 1 multispectral scanner. The major land systems can readily be delineated. Small patches of growing crops, having a high near-infrared reflectance, are red in this colour combination.

(a)

(b)

5. Photographic density slicing
(a) Infrared false-colour photograph of farmland and forestry plantations in south-western England.
(b) A combination of coloured density slices obtained from the same exposure by means of Agfacontour film. These enhance the details of crops and soil conditions and distinguish different tree types. (Courtesy Aerofilms Ltd.)

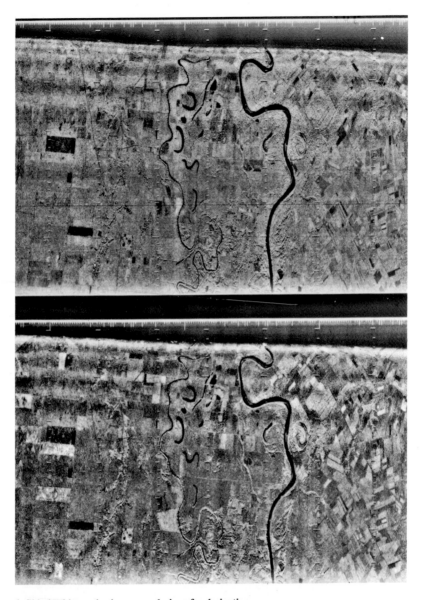

6. Side-looking radar imagery: choice of polarization
Dual strip image of an agricultural area in the southern United States. HV polarization recording in the top strip, HH below. The Horizontal banding is due to the
lobed aerial pattern of this real-aperture system. (Courtesy Westinghouse Electrical
Corpn.)

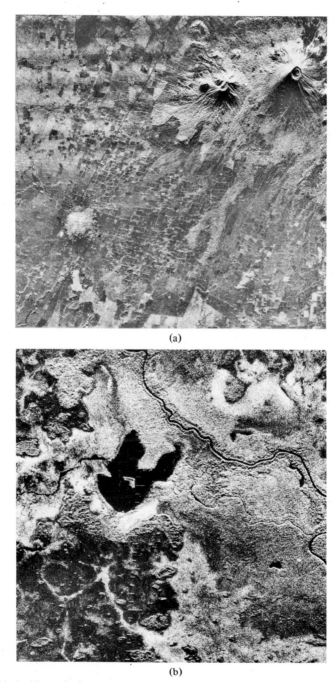

(a)

(b)

7. Side-looking radar imagery
(a) Imagery of an area in Nicaragua taken by a K-band real-aperture system, showing relief, natural vegetation, cultivation, and urban details.
(b) Enlarged section of a flat marshy area showing distinctive patterns of riverain forest and swamp savannah. Open water areas are black. (Courtesy Westinghouse Electrical Corpn.)

(a)

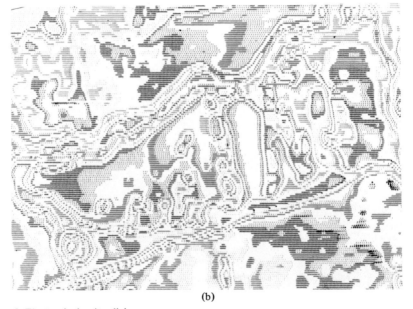

(b)

8. Electronic density slicing
(a) Thermal infrared linescan imagery (left), with a single black-and-white density slice (right) and combined colour-coded density slices (centre). Produced by a Densi-colour Video-Analysis system.
(b) Colour-coded density separations of part of an aerial photograph produced by a scanning microdensitometer (Joyce Loebl Isodensitracer).

between the need to cover a wide area frequently to monitor the dynamic weather situation and the load or 'data rate' on the telemetry link. As the quantity of data making up a picture increases in proportion to the square of the resolution of the sensor, the power available limits the quality of the picture that can be transmitted. As power is strictly limited on a small satellite, sensor resolutions are necessarily low.

The *Nimbus* satellites typically have a mean orbital height of just over 1000 km and an orbital period of revolution of just over 100 minutes. They are sun-synchronized, so that the ascending mode on the illuminated side of the earth occurs close to noon, local mean time, and assures good illumination for the imaging systems that operate in the visible band. The sensors provide a very wide picture or swath coverage (in some cases from horizon to horizon) and repeat the coverage of any one area several times a day, which is necessary for meterorological purposes. Because of its low resolution the imagery is useful only for special purposes in remote areas. For example, major geomorphic features can be distinguished, as well as ice and snow cover and tree-lines. Early examination of coverage of this kind did, however, serve to indicate the possibility of using automatic satellite pictures for purposes other than meteorology and oceanology.

ERTS *Landsat*

The Earth Resources Technology Satellite (ERTS 1), renamed *Landsat* 1 in January 1975, was the first satellite specifically designed to survey natural resources. It was launched by the National Aeronautics and Space Administration (NASA) on 23 July 1972 and was still in operation late in 1975. *Landsat* describes a near-polar orbit which traverses any point on the ground every 18 days. The height of the orbit is 494 nautical miles (946 km) and the satellite is sun-synchronous so that repeat coverage is always at approximately the same time of day. The orbit period is 103 minutes, giving 14 orbits per day (Fig. 7.1). The imaging swath of the sensors is 185 km wide.

On 22 January 1975 the first vehicle was joined by *Landsat* 2, an exact replica of the first satellite. *Landsat* 2 was, however, placed in an orbit at $90°$ to the inclination of *Landsat* 1 so that routine coverage of any area at 9-day intervals became possible using both satellites.

Both vehicles have two high-resolution automatic sensor systems: a four-channel multi-spectral scanner (MSS) and a three-camera return-beam vidicon system (RBV). The images are relayed by radio telemetry to ground stations initially at Fairbanks, Alaska, at Goldstone, California, and at Greenbelt, Maryland, where NASA processes the data to produce imagery. By 1975 additional stations had been or were

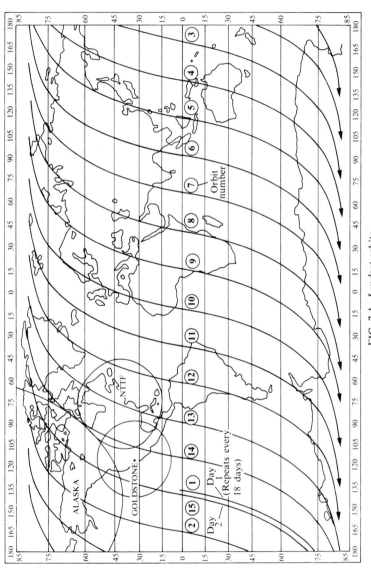

FIG. 7.1. *Landsat* orbits.

being set up in Canada, Brazil, Iran, Italy and Zaire, and were planned for a number of other countries.

When the satellite is out of direct line of sight of the receiving stations the image data is stored on magnetic tapes. These are played back when the satellite is again in a position to be interrogated by a ground station. Limited tape-recording capacities and some failures of parts of the equipment have restricted the amount of imagery obtained from outside north America. The supplementary stations should solve this problem. By late 1975, however, practically the whole world had been imaged by these satellites, many areas many times over. The problems of weather are such, however, that many areas did not have cloud-free coverage at any time.

The multispectral scanner carried by ERTS is similar in principle to the airborne systems described in Chapter 5. It records a continuous swath on the ground in a green, a red, and two near-infrared channels (Fig. 7.2a). Table 7.2 gives their performance specifications. The other imaging system of ERTS, the three return-beam vidicons, consists of three highly sensitive television cameras each filtered to give a separate picture in one of the wavebands listed in Table 7.3. The corresponding spectral curves are shown in Fig. 7.2b. When these three images are recombined, each with the appropriate filter, the result is an approximate simulation of a false-colour infrared Ektachrome photograph.

The RBV cameras take a succession of overlapping picture frames, covering the continuous 185-km swath of the MSS. These are video cameras of very high performance indeed, having over 4000 lines making up each picture frame compared to the 625 lines of the standard European television systems. A similar number of resolution elements along each line gives an overall number of over 16 million picture points per frame — equivalent to that of the MSS system.

The original intention was that the MSS would provide the most accurate spectral data and the RBV system would provide the more planimetrically correct imagery (see Table 7.4). In fact, owing to early difficulties with the RBV system, ERTS 1 provided very few pictures from this instrument compared to the number from the MSS. MSS picture data also proved in operation to have better planimetric (map) accuracy than had been expected.

Landsat imagery

At the ground receiving stations the picture data from the sensors on the *Landsat* vehicles are converted to digital form and recorded on videotapes which are passed on to the picture-processing centre. Here the image data from both the return-beam vidicon and the multi-

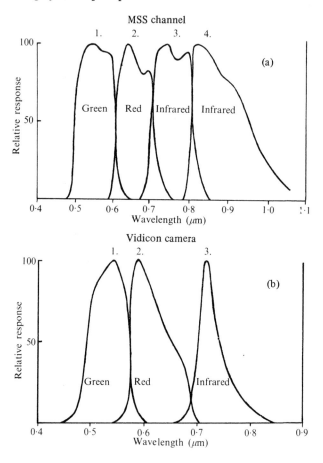

FIG. 7.2. Spectral curves for *Landsat* sensors.

spectral scanner systems are converted to film format by an electron beam recorder (EBR), which traces the data from the videotapes on to film at a density of over 4000 lines per frame. The pictures generated at this initial or 'bulk processing' stage are corrected for geometric distortion and for radiometric balance. Selected images can be 'precision-processed' by an alternative system to give pictures of much higher accuracy. This entails extensive computer processing to correct the images to measured ground control. The accuracies of the imagery generated by the alternative processing systems are shown in Table 7.4.

Table 7.2. ERTS/Landsat. Specifications of the multi-spectral scanner (MSS)

Spectral band	Wavelength (μm)	Detector
1 Blue–Green	0·5–0·6	
2 Yellow–Red	0·6–0·7	Photomultiplier tubes
3 Red–Near-infrared	0·7–0·8	
4 Near-infrared	0·8–1·1	Silicon photodiode

Width of swath coverage: 185 km
Instantaneous field of view of the optical system: 79 m

Table 7.3. ERTS/Landsat. Specifications of the return-beam vidicon system

Camera	Spectral band	Wavelength (μm)
1	Blue–Green	0·475–0·575
2	Green–Yellow	0·580–0·680
3	Red–Near-infrared	0·678–0·830

Frame coverage: 185 × 185 km (matching the MSS swath)
Resolution: 79 m

Table 7.4. Accuracy of ERTS imagery

	Multi-spectral scanner	Return-beam vidicon
Bulk-processed imagery		
Radiometric accuracy	± 3%	± 10%
Registration accuracy	115 m	325 m
Overall map accuracy	880 m	930 m
Precision-processed imagery		
Radiometric accuracy	± 4%	± 10%
Registration accuracy	120 m	110 m
Overall map accuracy	85 m	70 m

The bulk processing system produces imagery at a scale of 1:3 369 000 on 70-mm negatives; all other formats are generated from these. The most commonly used materials are 9½ -in prints at a scale of 1:1 000 000. These may be black and white printouts of one spectral band from the multispectral scanner or vidicon cameras, or a photographic combination of several. The most common combination presents the green, red, and near-infrared channels as blue, green, and red, which approximates to false-colour infrared Ektachrome air survey film (Pl. 4b). The resolution and positional accuracy of the bulk-processed images is high. For the most part it fulfils the accuracy specified for 1:250 000 mapping in the United States. Indeed, even enlarged bulk-processed imagery is considerably more accurate than the published 1:250 000 scale maps for some parts of the world.

Landsat imagery is freely and cheaply available for purchase from the United States Department of the Interior, through the EROS Data Center at Sioux Falls, South Dakota. The Bayswater Division of the United Kingdom Science Reference Library holds catalogues and microfilms of the *Landsat* imagery of the British Isles.

Skylab

One of the experimental programmes of the orbiting manned laboratory *Skylab* was to assess the use of remote sensing instruments for investigations into earth resources; it also produced extensive photography of the earth. The Earth Resources Experimental Package (EREP) that it carried included a six-channel multispectral camera pack, an 'earth terrain camera' of very long focal length, and a 13-channel multispectral scanner. Non-imaging microwave and infrared sensing equipment was also carried. Details are given in Appendix 4. Hand-held cameras could also be used from *Skylab* because it was a manned vehicle and exposed film could be brought back to earth so that much higher resolutions were possible than with *Landsat* (see Appendix 4).

Because of the limited periods during which the satellite was manned and the considerable pressure of other experimental work, *Skylab* produced much less coverage than the automatic imaging devices aboard ERTS 1 in its first year of operation. Coverage of high latitudes was not possible at all because of the low inclination of the orbit, which was limited to a track within latitudes 50° N and 50° S.

Black and white, colour, and false-colour photography from *Skylab* is also available in a number of formats from the EROS Data Center.

Uses of Landsat and Skylab imagery

Small-scale colour composite imagery or colour photography from ERTS or *Skylab* is now readily available for large parts of the world.

The novelty of *Landsat* and *Skylab* imagery, and encouragement from NASA, have prompted further investigations into its use.

Where aerial photography is not available, satellite imagery has been valuable as a substitute even though material on a larger scale at a higher resolution would have been preferred. It is increasingly seen to be of value in its own right as a tool for mapping at reconnaissance scales or as a complement to airborne photography or SLR imagery. ERTS imagery in standard format at scales from 1:1000 000 to 1:250 000 provides a good basis for the reconnaissance mapping of natural resources of all kinds, though the amount of supplementary ground information required can vary widely.

The basic geomorphological framework is usually well expressed on these materials. They thus suit the land-systems approach to soil mapping and they emphasize the relationships of geology and vegetation over wide areas. For example, it has been possible to produce geomorphological maps of the whole of Ethiopia with over 30 landform categories, using single-channel black and white imagery alone with a minimum of ground information. Although the geomorphic units provide an ideal framework for a reconnaissance soil map, a great deal of supplementary ground information is needed before the soils of all the units can be described with confidence. Areas of alluvial or colluvial soils, and the skeletal soil complexes of mountains or eroded lands, can be distinguished with considerable certainty, but it is not feasible to identify categories of zonal soils without ground information.

Much attention is being paid to 'stratified' approaches to natural resources mapping using *Landsat* image interpretation in combination with other survey and mapping techniques. An example of this is work begun in the Western Sudan in 1975 for a number of large-scale land development planning projects. The land systems, soil, and land capability mapping is based on using *Landsat* imagery both for interpretation and as base maps for compilation. Preliminary land systems analysis is first carried out on colour enlargements of *Landsat* frames. A limited number of exploratory field traverses selected from this material are then undertaken and one or two sample areas for each land system are investigated on the ground in detail. The boundaries from these areas are then extrapolated to refine the initial mapping. Final checking is carried out by visual observations from light aircraft flying systematic traverses. Hand-held camera shots may also be taken for additional verification.

This combined survey method has the advantage of allowing population, cattle, and wild animal counts and other inventory information to be taken at the same time. For the level of survey required — reconnaissance at 1:250 000 scale — the method is considerably more cost-effective than the conventional method of comprehensive field survey and small-scale aerial photo-interpretation.

Though there is much less *Skylab* photography than *Landsat* imagery it is being used for work of the same kind and for some purposes it is proving more useful because of its higher resolution. Where the photography from the earth terrain camera is of good quality it can be used to produce very satisfactory base maps at scales down to 1 : 100 000. The main drawbacks of *Skylab* cover are that for most areas only single strips were obtained rather than large areas of block cover and many of the exposures were obscured by cloud.

Though vegetation and moisture conditions are often ephemeral, the basic characteristics of soils are generally not liable to short-term change. For this reason frequent repetitive coverage is not as necessary for soil mapping studies as it is for monitoring crops and vegetation. A single set of good satellite cover will often be sufficient for reconnaissance mapping, though a number of satellite passes may be required to build up a complete set of pictures of good quality and free from cloud. The monitoring of soil erosion, or of soil moisture to determine irrigation requirements, may, however, require regularly repeated coverage.

Changes in vegetation or in soil moisture conditions may yield better indications of soil boundaries at some times of the year than at others. Examination of imagery from two or more dates may give much more information than a single set of cover. In the Southern Sudan differences in the dates of burning and the changing vigour of grassland during the wet and dry seasons are clearly shown on *Landsat* imagery taken at different times. The condition of the grasslands is directly related to the extent of seasonal flooding, a factor which to a large extent determines the basic characteristics of the soils in this area.

Future Space Programmes

The main development in satellite remote sensing of earth resources — already in the planning stage and due to be implemented in or by the early 1980s can be looked at under three headings. These are:—

i) *Global Coverage by Automatic Satellites*

There will be improved coverage by satellites of the *Landsat* type operating mostly in low, polar orbits to give frequently repeated coverage (every few days) of the whole world. The imagery will have better resolution — down to 30 metres and will be obtained in more spectral bands than at present.

The third of the *Landsat* series will be launched in 1977 and though similar in performance to the earlier vehicles it will have an additional — thermal infrared, channel. *Landsat* D (*Landsat* 4 on launching in

1981) will have considerably improved instruments. One possibility is for a pointable, very high resolution sensor with a relatively narrow field of view that can be directed at areas of interest by command from the ground. Another possibility suggested for the French Cameleon scanner is that the wavebands in which the images are recorded be similarly selected by ground control from a large number (11) of possibilities, depending on the requirements of the task in hand.

A number of countries are considering having their own earth resources satellites. These include France, Japan, West Germany and the Netherlands. India however is likely to be the first after the United States with its SEO — Satellite for Earth Observations — which will employ high resolution television cameras.

The European Space Agency (ESA) is also interested in earth resources satellites and has commissioned a number of definitions and feasibility studies for satellites of this sort. These include automatic satellites with side-looking radar and passive microwave mapping systems (see Appendix 5). The government of the Netherlands has undertaken similar studies working towards an 'Agrisat' vehicle, largely intended for crop and related studies.

(ii) *Selected Coverage from Manned Space Stations*

Very high resolution coverage, including conventional photography will be obtained from manned space stations like the European-American *Spacelab* and the USSR *Salyut*. *Spacelab* 1 is due for launching by the re-usable space shuttle in 1980 and these will be a continuing feature of the space programme from then on. Long duration manning of the Salyut stations has been continuing since 1974.

Continuous or frequently repeated synoptic coverage by manned stations may not be really practical, at least in the early stages. The orbits at first will tend to be at relatively low angles to the equator — the orbit of *Spacelab* 1 for instance will not extend beyond latitude 55° North or South. The pattern is likely to be of selected coverage obtained for specific purposes.

(iii) *Continuous Coverage by Geostationary Satellites*

Since 1974 geostationary satellites in orbits synchronised to the speed of revolution of the earth have been able to give continuous automatic picture coverage of most of the globe except for the polar regions. These are being used mainly for meteorological purposes and the imagery has resolutions of between 1 and 4 km only. Later versions of these will have increased resolution for their synoptic sensors and possibly much higher resolution steerable telescopic type sensors for selected local coverage — particularly for monitoring severe storms. It might well be that this sort of coverage could contribute to soil mapping by recording dynamic or changing features like soil moisture which could have a bearing on the physical nature of the soils.

Appendix 5 lists a number of planned and possible earth observation satellites. Military technology is inevitably well in advance of that applied to civil problems and can give some indication of future developments. The Russian Cosmos and the American Big Bird surveillance satellites carry very advanced imagery systems — probably with resolutions down to 30 cm in some cases.

It is probable that imagery from automatic satellites and recoverable films from manned missions will both have their place in mapping soils and other earth resources, particularly if the coverage remains readily and cheaply available. In general for cheap repeat coverage systematically obtained over long periods automatic satellites have the advantage. Photography from manned missions, however, has the greater potential for resolution and image quality.

8. Image enhancement and automatic image analysis

The rapid development of remote-sensing systems has led to a great deal of interest in automatic data-handling and image analysis. This is especially true of those techniques in which data are initially recorded in electronic form, for these require complex processing to make usable hard-copy images. The techniques used to reproduce imagery from electronic records can be adapted to enhance and extract selected features from the image records. This is the basis of automatic image analysis.

Between image acquisition and image analysis, however, there is often a *pre-processing* stage in which the images are spatially and spectrally rectified, and, for multi-spectral records, the separate channels of information are mutually registered. Pre-processing improves the imagery and translates it into more readily interpretable forms. It does not include analysis, which may vary from relatively simple enhancement procedures to procedures that entail complex computer processing.

Image enhancement

The object of enhancement is to extract information from the picture data that cannot be extracted by direct visual examination of the imagery in its normal format. The methods include relatively simple photographic and analogue electronic techniques and more advanced digital data-processing. With the aid of computers, quantities of image data that are far larger than a manual interpreter could assimilate can be processed very rapidly.

Photographic enhancement

Photographic *density slicing* is a simple method of selecting features from photographs or other hard-copy products. Any monochrome image consists of a continuous range of grey tones, or image densities, corresponding to the exposure levels from the different parts of the target. This range of densities or signal levels can be divided into a series of quantifiable steps. Consider a positive image produced by a single-channel thermal scanner. The lightest areas represent the warmest parts of the target, and the darkest the coolest. If the image is dissected into, say, ten separate density levels, each of these will represent a distinct temperature range and the boundaries between the areas of different density will be *isotherms* or temperature contours.

Photographic density slicing employs a specially prepared copy film with an emulsion containing silver bromide and silver chloride. When this film is exposed through an original negative, the most exposed (darkest) areas are registered by chemical development of the chloride grains and the least exposed by their physical development. Areas of medium exposure are only slightly developed, since they are below the threshold for chemical development of the chloride salts and they release bromide ions that inhibit the physical development of the chloride. The result of this is a single-tone image of the medium range of densities of the original image. This is a *density slice*. When a number of different slices are required the range of density in each is controlled by adjusting the development time and using filters to vary the light intensity of the exposures. Individual slices can be printed separately or in different colours for comparison (Pl. 5b). Density slices produced by photographic methods are usually relatively broad, though it has been claimed that as many as 28 separate density steps can be produced from a single image.

Separations of narrow density ranges can be obtained by super-imposing transparencies of slightly overlapping ranges — in effect subtracting one from another. Density slicing can be used to pick out the transition zones between adjacent images of differing density in order to produce a separation of their outlines resembling a line drawing. This procedure is known as *edge enhancement*. The same effect can also be obtained by other methods.

Unsharp masking can be used to produce edge-enhancement photographically. With this technique a positive transparency is exposed to a copy film in combination with a deliberately blurred negative of the same image. This cancels out most of the image, leaving only the edge detail.

Photographic image-enhancement is rapid and cheaper than electronic data-processing, but the processes depend upon chemical reactions that are difficult to control precisely and it can be difficult to obtain quantifiable results. A serious problem is that it is difficult to obtain even illumination over the whole image. As a result, a common fault with photographic density slices is darkening at the edges. This is due to the accentuation of edge effects owing to vignetting or uneven lighting during processing.

Video techniques

The simplest video enhancement systems use a television camera and display to look at hard-copy images or even at the actual scene. This procedure can be adapted to work from data received on magnetic tape in analogue or digital form. The picture on a television monitor can be density sliced electronically by selecting the signal levels that correspond to specific ranges of image response. With colour systems the individual slices can be assigned different colours. Up to 64 density levels can be discriminated by some video systems. With some systems images or sections of images can be selectively enlarged, edge enhancements can be made, and units of different density isolated and individually measured.

The pictures displayed by video systems, like those of multi-spectral photographic viewers, are generally best treated as ephemeral. Since the pictures are multi-variable there is little point in going to the expense of reproducing them as hard copy. Permanent records of the best combinations can be made by photographing the face of the screen, but the procedure is cumbersome and the results may have lost too much quality to be of value for interpretation. Interpretation is thus best carried out at the screen.

Video displays for interpretation suffer from the drawback that the image quality is degraded by the reduction in the number of lines of the display. These may be the 405, 550 or 625 lines of the British, United States, or European standard respectively. If the reduction to the line standard is not to reduce the resolution of an image, it can be viewed only in small enlarged segments. Television systems of higher resolution with 1000-line displays can be used but they are not generally available.

The use of computers

As high-resolution images of the type needed for soil surveys are essentially made up of very large numbers of 'bits' of spectral information (*resolution cells*), arranged in definite patterns, the use of computers, which can deal very rapidly with large amounts of numerical information, is a logical development in handling image data.

To be fed into a computer, image data must at some stage be in the form of magnetic tape recordings or their equivalent. As we have seen, electronic remote-sensing systems like linescanners may produce their records directly in this form, but for systems with hard-copy output, such as photographic cameras, the image has to be scanned by some form of electronic recording device to translate it into a computer-compatible format.

Tape recordings may be digital or analogue. Initially all electronic data, and for that matter photographs, are analogue in form but analogue tapes can be readily converted to digital formats. Any picture or tape can be digitized by translating the continuous scale of values of image density along a series of scan lines into discrete steps, to each one of which a numerical value is given. Each one of these steps is called a *picture point* or *pixel*. There are a number of instruments that can be used to convert hard-copy imagery to digital formats. These include mechanical scanning microdensitometers and flying-spot scanner (FSS) systems.

The principle of the scanning microdensitometer is that a very small spot of intense light is projected through, or reflected from, the image and its intensity, as modified by the image, is measured as a continuous line of fluctuating signal level, or sampled point by point along a series of lines scanning the image. The usual light sources are photodiodes but in some equipment laser sources are used. The image may be scanned on a flat table or on a revolving drum. Film is an excellent data store and high-resolution pictures contain very large quantities of data. A very small light spot is thus needed to 'read' the picture at its full resolution. The various types of equipment vary a great deal in spot size and speed of scanning, but in general drum scanners are very much faster than table-scanning instruments.

Flying-spot scanners work on the television camera principle, employing a beam from an 'electronic gun' to scan the image with high precision and rapidity.

Scanning microdensitometers and flying-spot scanners can both be used in an inverse mode, that is, instead of scanning data from a picture they can 'write' an image on to a film with their light source modulated by a data tape input.

Television cameras can be used for the input of image data by digitizing the picture within the system, but as already explained they are generally used only for low-resolution images.

These various items of equipment used as input and output devices are basic elements of complete image data handling systems, some of which are described below.

Computer handling and analysis of image data may be by analogue or digital processes. Because most computers are digital most work has been done on these. Analogue processing using special-purpose computers has the advantage of being faster than digital processing but there are difficulties with standardization and correlation of the different sets of data making up a complete survey.

The simplest type of image analysis by computer is density slicing, which can be achieved by 'thresholding' techniques. Many more levels can be extracted than by photographic methods: in some working systems there can be as many as 256. Edge-enhancement requires more complex programming, and even more complexity is required for image analysis proper, which requires automatic recognition of signatures.

Signatures and image analysis

The densities of a picture point of a multi-spectral image in each of its channels constitute the *spectral signature* at that point. If these signatures, in numerical form, can be recognized, classified, and printed out by a computer, an automatic mapping system is possible.

The human eye can successfully integrate only three or four channels of spectral information, since in normal colour vision there are only three primary colours; blue, green, and red. The relationship between any two of these can be represented as a scatter diagram. It is relatively easy to extract from a three-layer colour photograph a two-dimensional plot which records the intensity at a number of selected points in the red waveband compared to the green. A plot incorporating the blue information at the same points would need to be reproduced in three dimensions. Computers, on the other hand, can deal with recordings in any number of spectral bands, working in effect in 'n-dimensional space' (n being the number of separate channels of data in each signature). Only a computer can fully analyse data from, say, a 24-channel multi-spectral scanner.

The multiplication of spectral channels inevitably increases the load on the computer. For example, to deal with all the pixels on a single frame of a *Landsat* image (approximately 4000 × 4000 in each spectral channel) a computer has to manipulate 16 million numbers. Only the largest computers can handle such quantities of data, but the problem can be eased by processing images of this type in segments, or by pre-processing the data to reduce the quantity. Image data can be *bulked* by combining or averaging groups of pixels. This procedure, however, reduces the resolution of the final image.

Once signatures can be recognized it is theoretically possible to convert multispectral data into maps in which each picture element is assigned to a specific class. This is done by matching the spectral signatures with the range of signatures determined for the class. These class groupings may be determined by 'ground truth' spectral measurement or deduced from natural groupings which can be related to mappable ground features.

There are several approaches to computer analysis of multispectral data. Four of these are: clustering algorithms, likelihood ratio algorithms, 'table look-up' procedures, and ratioing (Van Vleck *et al.* 1973).

Clustering algorithms

Clustering algorithms take the picture data points of the plot in n-dimensional space and look for natural clusters based on proximity relationships. Once clusters are recognized, each picture point can be labelled for the cluster to which it belongs and printed out separately or in a distinctive colour code. Ambiguous points falling between clusters are allocated by the computer to the nearest of the clusters. The technique works best when all picture elements fall into well-differentiated signature classes; serious difficulties arise when clusters overlap. In such circumstances the regions of ambiguity have to be labelled separately as 'unresolved', and additional data input is thus required.

This technique creates its own classification system by recognizing natural groupings and it does not, in the first instance, need ground truth data inputs. It is an *unsupervised learning* approach and its usefulness depends on whether the classes it erects have any practical significance. Having produced a classification and a map, its users have then to ascertain if the mapping units represent significant divisions.

Likelihood ratio algorithms

Likelihood ratio algorithms are used in a 'supervised learning' approach in which the spectral characteristics of 'ground truth' samples of the units to be mapped are first measured and used to determine the *spectral classes* to which they correspond. An '*n*-dimensional decision rule' is then used to assign every other picture point in the imagery to the class to which it has the maximum likelihood of belonging according to its signature.

This definition of the algorithm or *training* that is introduced before the analysis of the image data can be applied to both analog or digital processing. The SPARC analog computer of the Environmental Research Institute of Michigan uses this technique with up to 12 channels of multi-spectral scanner data. Analog computers have a higher potential throughput than digital machines, most of which do not have the capacity to handle inputs from large numbers of channels. Training processes are, however, more laborious with analog machines. Hybrid analog–digital computer systems may prove eventually to be the most effective solution for this kind of work.

Table look-up procedures

Table look-up procedures are essentially a digital approach, which makes use of the storage capacity of the computer rather than its capacity for high-speed calculation. With this procedure the distribution or range of variation of the spectral signature in each class is recorded from ground truth, or training, sites and the information is stored in the computer. In processing the multi-spectral survey data each picture point or pixel is allotted to a class by directly matching it to the stored data. This technique has the advantage that the procedure for identifying the class to which any pixel belongs is a simple 'fetch' operation rather than a more complex calculation. It does, however, require a very large storage capacity and in the present state of the art it can be performed with no more than four channels of data. It may be possible in some circumstances to reduce the amount of storage required by determining in advance what data are really needed for classification. For instance, two channels of data from a multi-spectral set may have a constant relationship to each other in more than one spectral class and if they are not critical in class discrimination they can be discarded as redundant.

Of these three methods the maximum likelihood approach using an analogue computer is considered the fastest, followed by table look-up procedures with a digital computer. Both these are faster than using the maximum likelihood approach or clustering algorithms on a digital machine.

Ratioing

Ratioing is a technique for image enhancement rather than for automatic signature analysis. It is used to emphasize subtle tonal variations. The computer calculates the ratios of the image intensities in two or three spectral channels for each picture point. These ratios indicate the differences in spectral response between channels; if two intensities are identical they are cancelled out. Only the differences are printed out as a separate set of values.

Automatic mapping

Automatic mapping from image data depends on some form of spectral analysis, but the spectral classes must correspond to relevant mapping units. If discrete signatures or ranges of signatures cannot be derived from specific objects, areas, or targets, they cannot be mapped automatically in this way. However successful automatic mapping procedures might become, ground-truth information will always be required at some point. To take the simplest possible illustration, an instruction to separate green fields with growing crops from brown fields without crops is a ground-truth input. In practice most examples are far more complex than this and require more subtle discriminations. A field of wheat, for instance, may be green overall but it is not uniformly of one tone and an adjacent field of beans may be a different shade of green. However, despite the variation in each crop the overall signatures of wheat and beans can be readily distinguished from each other by a computer.

Virtually all natural mapping units have some variability in their spectral response. In general, the greater the complexity the greater is the likelihood of ambiguities arising in automatic classification. The introduction of more spectral channel inputs will help to reduce the number of ambiguous points in an image, but the larger the number of channels used the larger the number of interactions required in the computer and the greater the cost of computing.

A complicating factor in the use of automatic mapping techniques is the variation in spectral signatures for similar classes over distance. With any but the simplest types of discrimination the validity of any one set of ground-truth data has a limited range. Current work at Purdue University in the United States with multi-spectral scanner data indicates that signatures are reasonably valid only within a radius of between 8 and 80 km.

Most work on automatic image analysis has so far been on crop mapping, with some mapping of broad groups of natural vegetation and soils. Military work on target identification has mostly been concentrated on single-channel imagery.

The dissemination by NASA of image data from ERTS 1 on magnetic tape has generated a great deal of interest, particularly in university geography departments working with computers. Most of this work is still experimental and it is difficult to identify practical applications. The great wealth of literature is mainly on general methods and results rather than on working systems for practical problems. Soils are more difficult targets than crops, for they often grade from one type to another without sharply defined boundaries. In addition, boundaries are often obscured by vegetation and by surface modifications such as crusts.

As observed signatures in any case relate to the immediate surface rather than to the complete profile, the chances of a high degree of correlation between soils and spectral responses are not as good as for vegetation types. The 'groupings of soil mapping units' that some experimenters have identified by automatic mapping correspond more to land management types than to classes defined on purely pedological factors (Cipra *et al.* 1972).

Image analysis systems

A number of image analysis systems have been developed by different organisations, using a variety of equipments. Some of the different elements of complete systems have been described earlier in this chapter, namely input and output devices. The make-up of an image-analysis system is essentially modular, composed of the elements illustrated in Fig. 8.1.

Analysis of image data can be carried out on general-purpose computers using standard line printer equipment to display the results

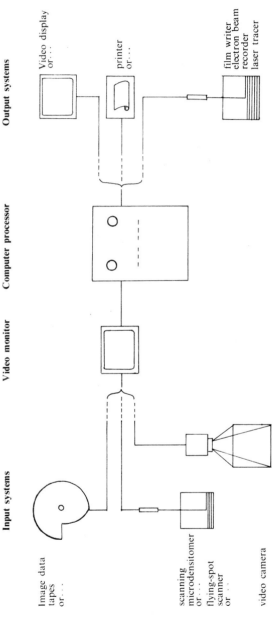

FIG. 8.1. Components of an image analysis system.

graphically. The more advanced work requires specially engineered systems and units. An example is the *I*mage *D*iscrimination, *E*nhancement, and *C*ombination *S*ystem (IDECS) at the University of Kansas.

IDECS is an analogue–digital system that can accept multiple image inputs. These are normally phtographic transparencies which are converted to electronic data by three synchronized flying-spot scanners. Non-transparent images can, however, be fed in by video cameras. The input systems are linked to a *core processor* which is based on a PDP–15/20 computer. Density-sliced multi-spectral combinations and other analysed images are displayed on television screens, either black and white or colour. The video camera inputs are of much lower resolution than those from the flying-spot scanners and are generally used to feed in the base maps against which the images being interpreted are to be registered for correct orientation. As well as providing the displays for the interpreter, the video displays have variable sampling grids to take off processed data for statistical analysis of the images. Although originally developed for use with hard-copy side-looking radar imagery, the system can be adapted to accept image data directly from analogue or digital tape. (Anderson et al 1972).

Other image analysis systems include the LARS system at Purdue University and the Image 100 System produced by the General Electric Company in the United States, and a system known as CERES developed by the Plessey Company in the United Kingdom. The emphasis in most developing systems is on modular designs with interchangeable units to accept varied inputs – hard copy or data tape – and to produce varied outputs – real-time displays or high-quality permanent picture records. Another aspect that is receiving much attention is man-machine interaction. The object here is to facilitate data-handling, particularly mutual registration of different inputs, and to help to eliminate ambiguities and discrepancies in processing.

Summary

The main advantages of computer and other electronic analysis of image data can be summed up as follows: more subtle differentiations can be discerned than is possible with the unaided eye; more sets of spectral data can be assimilated than by manual methods; very large quantities of data can be processed very rapidly. The first of these may be an advantage in soil mapping, but most soil properties of interest lie below the surface and surface sensing, however sophisticated, can give only partial answers. Nor is rapid processing of large quantities of data so important a requirement in mapping soils or any other features that are not subject to rapid change.

For the immediate future, it appears likely that the main value to soil survey of the computer processing of imagery will be in pre-processing to provide the interpreter with pictures of better quality. Simpler image-enhancement equipment is useful for 'quick-look' appraisals of imagery for reconnaissance surveys or for selecting data formats for interpretation or for illustrations.

Selected References

American Society of Photogrammetry (1960). *Manual of photographic interpretation*. George Banta Co., Menasha, Wisconsin.

American Society of Photogrammetry (1966). *Manual of photogrammetry*. George Banta Co., Menasha, Wisconsin.

American Society of Photogrammetry (1968). *Manual of colour air photography*. George Banta Co., Menasha, Wisconsin.

American Society of Photogrammetry and Society of Photographic Scientists and Engineers (1969). *New horizons in colour aerial photography*. Seminar Proceedings. Washington, D.C.

American Society of Photogrammetry (1973). *Management and Utilization of Remote Sensing Data*. Symposium Proceedings, Sioux Falls. (Session IIIc, Digital techniques; Session IV b, Non-digital and Man-assisted techniques).

Anderson, P.N., *et al.* (1972). Image processing with a hybrid system: the IDECS. University of Kansas Publications, *Bulletin of Engineering* No. 64.

Barr D.J. and Miles R.D. (1970). SLAR imagery and site selection. *Photogrammetric Engineering* 36 (2), 1155−70.

Bawden, M.G., (1967). Application of aerial photography in land systems mapping. *Photogrammetric Record,* 5 (30), 461−4.

Brock G.C. (1967). *The physical aspects of aerial photography*. Dover Publications, New York.

Carrol D.M. (1973). Remote sensing techniques and their application to soil science. *Soils and Fertilizers,* 36 (7), 259−66.

Cipra, J.E., Swain P.H., Gill, J.H., Baumgardner, M.F., and Kristof, S.J. (1972). Definition of spectrally separable class for soil survey research. *Proceedings of the Eighth International Symposium on Remote Sensing of the Environment, Ann Arbor, Michigan*, pp. 765−70. University of Michigan.

Colvocoresses, A.P. (1974). Remote sensing platforms. *Geological Survey Circular* 693. United States Geological Survey, Reston, Virginia.

Condit, H.T. (1970). The spectral reflectance of American soils. *Photogrammetric Engineering*, 36 (9), 955−66.

Crea, W. (1973). Agriculture, forestry, range resources. In *Symposium on significant results obtained from earth Resources Technology Satellite−1*, Vol. 3, Discipline summary reports, pp. 1−14. National Aeronautics and Space Administration.

de Godoy, R.C. and Van Roessel, J.W. (1972). Semi-controlled SLAR mosaics for Project Radam. *Proceedings of the American Society of Photogrammetry*, 39th Annual Meeting.

Goosen, D. (1967). Aerial photo interpretation in soil survey. *Soils Bulletin* No. 6. Food and Agricultural Organization of the United Nations, Rome.

Heller, R.C. (1970). Imaging with photographic sensors. In *Remote Sensing with special reference to agriculture and forestry*. National Academy of Sciences, Washington, D.C.

Holter M.R. (1970). Imaging with non-photographic sensors. In *Remote sensing with special reference to agriculture and forestry*. National Academy of Sciences, Washington, D.C.

Howard, G.A. 1973. Passive remote sensing of natural surfaces by reflective techniques. *Geoexploration*, 2 (3).

Innes, R.B. (1972). Controlled quality images from synthetic-aperture radar data. *Seminar on Operational Remote Sensing, Houston*. American Society of Photogrammetry, Washington, D.C.

Kodak Ltd. (1972). Starting professional colour printing. Kodak Ltd. *Professional Publication* KLD−21.

Kristof, S.J., Zachary, A.C., and Cipra, J.E. (1971). Mapping soil types from multi-spectral scanner data. *Proceedings of Ann Arbor Symposium on Remote Sensing*, 1971, pp. 2095−108.

Lyon, R.P.J. (1965). Analysis of rocks by spectral infra-red emission. *Economic Geology*, **60**, 715−36.

MacDonald, H.C. and Waite, W.P. (1971). Soil moisture detection with imaging radars. *Water Resources Research*, **7** (1), 100−10.

Marshal, A. (1968). Infra-red colour photography. *Science Journal*, January 1968, 45−51.

Mott, P.G. (1966). Colour aerial Photography in Practice and Application. *Photogrammetric Record*, **5** (28), 221−37.

Mitchell, J. (ed.) (1955). *The Ilford manual of photography*. Ilford Ltd., London.

National Aeronautics and Space Administration (1972). *Earth Resources Technology Satellite−1*. Symposium Proceedings, September 1972. Goddard Space Flight Center, Maryland.—— (1973) *Symposium on significant results obtained from Earth Resources Technology Satellite−1*, March 1973. Goddard Space Flight Center, Maryland.

Parry, D.E. (1974). Natural resources evaluation of ERTS−1. Imagery of the Central Afar region in Ethiopia. *Photogrammetric Record*, **8** (43), 64−80.

Parry, J.T., Cowan, W.R., and Heginbottom, J.A. (1969). Soil studies using colour photos. *Photogrammetric Engineering* **35**, 44−56.

Pesce, A. (1968). *Gemini Space Photographs of Libya and Tibesti*. Petroleum Exploration Society of Libya, Tripoli.

Piech, K.R., and Walker, J.E. (1974). Interpretation of soils. *Photogrammetric Engineering*, **40** (1), 87−94.

Pomerening, J.A., and Clive, M.G. (1953). The accuracy of soil maps produced by various methods that use photo-interpretation. *Photogrammetric Engineering*, **19** (5), 809−17.

Sabatini, R.R., Rabshevsky, G.A., and Sissala, J.E. (1971). *Nimbus earth resources observations.* Allied Research Associates Inc., Concord, Mass.

Sabins, F.F. (1973). Flight planning for thermal IR. *Photogrammetric Engineering,* 39 (1), 49—58.

— (1973). Recording and processing thermal IR. Imagery. *Photogrammetric Engineering,* 39 (8), 839—44.

Stoner, E.R., and Horvath, E. (1971). The effect of cultural practices on multi-spectral responses from surface soil. *Photogrammetric Engineering,* 39 (8), 2109—13.

Swanlund, G. (1972). Hybrid techniques for automatic imagery interpretation. *Seminar on Operational Remote Sensing, Houston.* American Society of Photogrammetry, Washington.

Stobbs, A.R. (1970). Soil survey procedures for development purposes. In *New possibilities and techniques for land use and related surveys.* Geographical Publications Ltd.

Van Genderen, J.C. (1972). An integrated resources survey using orbital imagery – an example from south-east Spain. *Proceedings of Ann Arbor Symposium on Remote Sensing,* 1972, pp. 117—35.

Van Vleck, E.M., Sinclair, K.F., Pitts, I.W., and Slye, R.E. (1973). *Earth resources ground data handling systems for the 1980s.* NASA Technical Memorandum NASA TMX—62, 240.

Vincent, R.K., and Thomas, F. (1971). Discrimination of basic silicate rocks by recognition maps processed from aerial infra-red records. *Seventh International Symposium on Remote Sensing,* pp. 247—52. University of Michigan.

Viglione, S.S. (1972). Digital data processing and analysis. *Seminar on Operational Remote Sensing, Houston.* American Society of Photogrammetry, Washington.

Wilson, C.L. (1972). Multi-spectral scanner applications. *Seminar on Operational Remote Sensing, Houston.* American Society of Photogrammetry, Washington.

Appendix 1 Members of the British Air Surveys Association

BKS Surveys Ltd.
Ballycaim Road, Coleraine BT51 3HZ, County Londonderry,
Northern Ireland.

Fairey Surveys Ltd.
Reform Road, Maidenhead, Berkshire SL6 8BU, England

Hunting Surveys Ltd.
Elstree Way, Borehamwood, Herts WD6 ISB, England.

Meridian Airmaps Ltd.
Commerce Way, Lancing, Sussex, England.

Appendix 2   Aerial Photography libraries in the United Kingdom

Central Register of Air Photography of England & Wales,
Air Photographs Unit, Department of the Environment, PDPS5,
Prince Consort House, Albert Embankment, London S.E.1.

Central Register and Library of Aerial Photographs of Scotland,
Scottish Development Department, Graphics Group, Air Photographs
Unit, York Buildings, Queen Street, Edinburgh EH2 1 HY.

Note: These are public collections. The air survey companies listed in Appendix 1
also have their own collections.

Appendix 3 Photography from manned satellites

Gemini V	21−8 August 1965
Gemini VII	4−18 December 1965
Gemini IX	3−6 June 1966
Gemini XI	12−15 September 1966
Apollo VI	4 April 1968
Apollo VII	11−12 October 1968
Apollo IX	8−12 March 1969
Skylab 2*	25 May−22 June 1973
Skylab 3	28 July−25 September 1973
Skylab 4	16 November 1973−8 February 1974

* Periods during which the vehicle was manned

Appendix 4 Skylab: EREP (Earth Resources Experimental Package)

(1) Multi-spectral photographic camera (MPC)

Six cameras with matched lenses.

Focal length: 152 mm (6 in)
Field of view: 21·2°
Area of coverage: 160 × 160 km
Film format: 70 mm
Acquisition scale: 1:2 850 000

Film−filter combinations

Filter bandwidth (µm)	Film	Resolution
0·5−0·6	Pan-X black and white	30 m
0·6−0·7	Pan-X black and white	28 m
0·7−0·8	IR black and white	68 m
0·8−0·9	IR black and white	68 m
0·5−0·88	IR colour	57 m
0·4−0·7	High resolution colour	24 m

(2) Earth terrain camera (ETC)

High-resolution camera with manually interchangeable film and filters

Focal length: 460 mm
Area of coverage: 108 × 108 km
Film format: 128 mm (5 in).
Frame overlap: 85 per cent
Acquisition scale: 1:950 000

Film−filter combinations

Filter bandwidth (µm)	Film	Resolution
0·4−0·7	High-resolution colour	15·3 m
0·5−0·7	High-definition black and white	15·3 m
0·5−0·88	IR colour	15·3 m

(3) Multispectral scanner (MSS)

Thirteen-channel scanner

Aperture: 44·5 mm
Field of view: 0·182 milliradians square (79 m) from 428 km.
Swath coverage: 67 km
Scan rate: 95 rev/s.

Channel	Bandwidth(μm)	
1	0·41 – 0·46	} Blue
2	0·46 – 0·51	
3	0·52 – 0·56	} Green
4	0·56 – 0·61	
5	0·62 – 0·67	} Red
6	0·68 – 0·76	
7	0·78 – 0·88	
8	0·98 – 1·08	} Near-infrared
9	1·09 – 1·19	
10	1·20 – 1·30	
11	1·55 – 1·75	
12	2·10 – 2·35	} Middle infrared
13	10·2 – 12·50	

Detectors: mercury cadmium telluride array.

(4) Infrared spectrometer

A filter wheel spectrometer with a narrow field of view operating in selective wavebands from the ultraviolet to the thermal infrared.

(5) Microwave radiometer/scatterometer and altimeter

Using a mechanical scanning reflector for active and passive microwave measurements in the frequency 13·9 GHz.

(6) L-band radiometer

For passive microwave measurements. Centre bandwidth: 1·4125 GHz. (wavelength: 212·4 mm).

Appendix 5 Earth observations satellites
(including planned and possible programmes)

Satellite/Programme

1961	*Mercury*−1	First of a manned series. Hand-held cameras used.
1965	*Gemini*−3	First manned vehicle of the series. Hand-held camera used.
1968	*Apollo*−7	First manned vehicle of the series. Cameras and other instruments used.
1972	ERTS−1 (Landsat−1)	Automatic Earth Resources Technology Satellite.
1973	*Skylab*	Manned space station. Cameras and other sensors used.
1974	*Salyut* 3 and 4	Manned space stations. Cameras and other sensors used.
1975	*Landsat*−2	Automatic Earth Resources Satellite.
1977	HCMM	Heat Capacity Mapping Mission for low-resolution world-wide infrared mapping.
1977	*Landsat*−C	Automatic Earth Resources Satellite continuing the programme.
1978	SEO	Indian Earth Observations Satellite
1980?	ERSS	Earth Resources Satellite: France.
1980?	ERSS	Earth Resources Satellite: Japan.
1980	ARTISS	Agricultural Monitoring Satellite: Netherlands
1980	*Spacelab*−1	Carrying an Earth Resources Survey Package: ESA.
1981	*Landsat*−D	Advanced Automatic Earth Resources Satellite with a sensor package with a considerably improved performance.
1981	SEOS	Synchronous Earth Observations Satellite.
1980s	SARSAT	Synthetic Aperture Radar Satellite: ESA.
1980s	PAMIRASAT	Passive Microwave Radiometer Satellite: ESA.

Subject Index

Principal references are shown in bold type.